Praise for *Don't Even Think About It*

"The science of climate change is easy: burning fossil fuels creates greenhouse gasses that are warming our world. George Marshall reminds us about the hard part: connecting the wellhead to the tailpipe in people's minds as soon as possible." —Bill Nye

"George Marshall is one of the most interesting, challenging, and original thinkers on the psychology of our collective climate denial. If his advice were heeded, we might just have the courage to look unblinkingly at this existential crisis, and then to act." —Naomi Klein, author of *This Changes Everything* and *The Shock Doctrine*

"[Marshall] offers advice on confronting climate change head on, stepping away from Green Guilt, and putting potentially world-saving policies into action." —*The Boston Globe*

"Engaging . . . His work is a much needed kick in the pants for policymakers, grassroots environmentalists, and the public to induce us to develop effective motivational tools to help us take action to face the reality of climate change before it's too late." —*Booklist*

"Absorbing, all-embracing, immensely readable." —*Climate News Network*

"Articulate, well-researched . . . Low on jargon and high on story . . . Read [this book] even if the words 'climate change' make you fall asleep." —*The Independent on Sunday*

"Makes clear why we continue down a dangerous path of increasing climate disruption, even when attractive, hospitable, alternative paths are available." —James Hansen, former director of NASA Goddard Institute for Space Studies, author of *Storms of My Grandchildren*

"This is not a book to read and put away—but one that merits returning to and engaging with intellectually." —*Daily Kos*

"[A] lively, nonpolemical account . . . An insightful, often discouraging look at why climate control advocates have failed to get their message across and what they should do. Much of Marshall's advice is counterintuitive (e.g., drop the apocalyptic rhetoric), but it rings true." —*Kirkus Reviews*

"Intriguing . . . Offers many answers." —*London Review of Books*

"A real soul-searching challenge for us all. Marshall illuminates the path to embarking on a heroic quest for a just and equitable world. A sobering, yet hopeful book." —Frank DiSalvo, director of the Atkinson Center for a Sustainable Future, Cornell University

"Enlightening." —*Publishers Weekly*

"Expansive and engaging . . . *Don't Even Think About It* brings an important perspective—that of the social sciences—to the debate over climate change and presents that perspective in an accessible and engaging way. The climate debate, Marshall demonstrates, is no longer about carbon dioxide and temperature-change models. It's about biases, values, and ideology." —*Stanford Social Innovation Review*

"Fantastic." —*Grist*

"Essential reading for everyone interested in communicating the science of climate change and its urgent policy implications." —*Critical Angle*

"Having cast gloom in all directions, Marshall offers some solutions. He suggests ways of changing the message to make the threat seem imminent and prompt people to take action . . . If the experts interviewed in this book are correct, then we are wired to allow our own destruction, and the fossil-fuel industry is so skilled at undermining the science and the political process that they will make sure we do. Let's hope the experts are wrong." —*Resurgence & Ecologist*

DON'T EVEN THINK ABOUT IT

DON'T EVEN THINK ABOUT IT

WHY OUR BRAINS ARE WIRED TO IGNORE CLIMATE CHANGE

GEORGE MARSHALL

BLOOMSBURY

LONDON · OXFORD · NEW YORK · NEW DELHI · SYDNEY

To Annie, Ned, and Elsa

Bloomsbury USA
An imprint of Bloomsbury Publishing Plc

1385 Broadway	50 Bedford Square
New York	London
NY 10018	WC1B 3DP
USA	UK

www.bloomsbury.com

BLOOMSBURY and the Diana logo are trademarks of Bloomsbury Publishing Plc

First published 2014
This paperback edition published 2015

ISBN: HB: 978-1-62040-133-0
PB: 978-1-63286-102-3
ePub: 978-1-62040-134-7

LIBRARY OF CONGRESS CATALOGING-IN-PUBLICATION DATA HAS BEEN APPLIED FOR.

2 4 6 8 10 9 7 5 3 1

Typeset by Hewer Text UK Ltd, Edinburgh
Printed and bound in the U.S.A. by Thomson-Shore Inc., Dexter, Michigan

To find out more about our authors and books visit www.bloomsbury.com.
Here you will find extracts, author interviews, details of forthcoming events,
and the option to sign up for our newsletters.

Bloomsbury books may be purchased for business or promotional use.
For information on bulk purchases please contact Macmillan Corporate and
Premium Sales Department at specialmarkets@macmillan.com.

Contents

1

Questions

*In 1942 the Polish resistance fighter Jan Karski gave eye witness
testimony to the Supreme Court judge Felix Frankfurter of the
clearing of the Warsaw Ghetto and the systematic murder of Polish
Jews in the Belzec concentration camp. Listening to him, Frankfurter,
himself a Jew, and one of the outstanding legal minds of his generation, replied, "I must be frank. I am unable to believe him." He
added: "I did not say this young man is lying. I said I am unable to
believe him. There is a difference."*

WHAT EXPLAINS OUR ABILITY TO separate what we know from what we
believe, to put aside the things that seem too painful to accept? How is it
possible, when presented with overwhelming evidence, even the evidence
of our own eyes, that we can deliberately ignore something—while being
entirely aware that this is what we are doing?

These questions have fascinated me for all the years I have been working on climate change*. They are what drew me to write this book and to
spend years speaking with the world's leading experts in psychology,
economics, the perception of risk, linguistics, cultural anthropology, and
evolutionary psychology, not to mention hundreds of non-experts—ordinary people I have encountered on the way.

At each step in this journey, as I tried to understand how we make

---

* Yes, climate is always changing, but here I am following the international legal definition
as being "attributed directly or indirectly to human activity that alters the composition of
the global atmosphere and which is in addition to natural climate variability observed over
comparable time periods."

sense of this issue, I uncovered other intriguing anomalies and paradoxes demanding explanation:

- Why do the victims of flooding, drought, and severe storms become *less* willing to talk about climate change or even accept that it is real?
- Why are people who say that climate change is too uncertain to believe more easily convinced of the imminent dangers of terrorist attacks, asteroid strikes, or an alien invasion?
- Why have scientists, normally the most trusted professionals in our society, become distrusted, hated, and the targets for violent abuse?
- Why is America's most prestigious science museum telling more than a million people a year that climate change is a natural cycle and that we can grow new organs to adapt to it?
- Why are science fiction fans, of all people, so unwilling to imagine what the future might really be like?
- Why does having children make people less concerned about climate change?
- How did a rational policy negotiation become a debating slam to be won by the wittiest and most aggressive player?
- Why can stories based on myths and lies become so compelling that a president prefers to take his climate science advice from a bestselling thriller writer rather than the National Academy of Sciences?
- And why *is* an oil company so much more worried about the threats posed by its slippery floors than the threats posed by its products?

Through asking these questions I have come to see climate change in an entirely new light: not as a media battle of science versus vested interests or truth versus fiction, but as the ultimate challenge to our ability to make sense of the world around us. More than any other issue it exposes the deepest workings of our minds, and shows our extraordinary and innate talent for seeing only what we want to see and disregarding what we would prefer not to know.

I work for a small educational charity, advising other nonprofits, governments, and businesses on how they can better talk about a subject that most people don't really want to talk about at all. I spend most of my working life with people like myself—concerned, well informed, liberal minded environmentalists—so it was a pleasant surprise, while writing this book, to discover I often learned the most from the people who are entirely different from me.

Talking to Texan Tea Partiers led me to ask why we climate communicators have so singularly failed to connect with their concerns. Speaking to evangelical leaders made me question the boundaries between belief and knowledge. I have even enjoyed meeting the people whose life work, to which they apply great dedication and creativity, is to undermine my own life work.

So I do not seek to attack the people who do not believe in climate change. I am interested in how they reach those conclusions, and I am just as interested in how believers reach and hold theirs. I am convinced that the real answers to my questions do not lie in the things that drive us apart so much as in the things we all share: our common psychology, our perception of risk, and our deepest instincts to defend our family and tribe.

These ancient skills are not serving us well. In this book I argue that climate change contains none of the clear signals that we require to mobilize our inbuilt sense of threat and that it is remarkably and dangerously open to misinterpretation.

I find that everyone, experts and non-experts alike, converts climate change into stories that embody their own values, assumptions, and prejudices. I describe how these stories can come to take on a life of their own, following their own rules, evolving and gaining authority as they pass between people.

I suggest that the most pervasive narrative of all is the one that is not voiced: the collective social norm of silence. This response to climate change is all too similar to that other great taboo, death, and I suggest that they may have far more in common than we want to admit.

I argue that accepting climate change requires far more than reading the right books, watching the right documentaries, or ticking off a checklist of well-meaning behaviors: It requires conviction, and this is difficult to form and even harder to maintain. It took me many years to reach my own personal conviction that climate change is real and a deadly serious threat to everything I hold dear. This is not easy knowledge to hold, and in my darker moments I feel a deep sense of dread. I too have learned to keep that worry on one side: knowing that the threat is real, yet actively choosing not to feel it.

So I have come to realize that I cannot answer my questions by looking too long at the thing that causes this anxiety. There are no graphs, data sets, or complex statistics in this book, and I leave all discussion of possible climate impacts until a final postscript at the very end. This is, I am certain, the right way around. In the end, all of the computer models, scientific predictions, and economic scenarios are constructed around the most important and uncertain variable of all: whether our collective choice will be to accept or to deny what the science is telling us. And this, I hope you will find, is an endlessly disturbing, engrossing, and intriguing question.

2

We'll Deal with That Lofty Stuff Some Other Day

Why Disaster Victims Do Not Want to Talk About Climate Change

WENDY ESCOBAR REMEMBERS FEELING SLIGHTLY nervous as she set off with her children to pick up groceries and saw the distant spiral of smoke on the horizon. But she says she never, ever, could have anticipated the speed or intensity of the disaster that followed. By the time she returned an hour later, the police had erected barricades across Texas State Highway 21. She had nothing but the clothes on her back; her daughter, she recalls, was still in her slippers. Two weeks later, when the road was finally reopened, the only family possession she could find in the ashes of the house was her great-grandfather's Purple Heart medal. It was melted almost beyond recognition.

The Bastrop wildfire of October 2011 was exceptional by any standards. Supercharged by thirty-mile-per-hour winds during a period with the lowest annual rainfall ever recorded, it killed two people, burned fifty-four square miles of forest, and could be seen from outer space. It destroyed 1,600 houses; ten times more than any previous wildfire in Texan history.

What was curious, though, was that, when I visited Bastrop a year later not one person, in a string of formal interviews, could recall for me a single conversation in which they had discussed climate change as a potential cause of the drought or the fire.

As one would expect in rural Texas, many people were unconvinced about the issue: many people, but not all. Wendy Escobar, for example, who laughed about "us being all rednecks out here," is an intelligent and thoughtful woman who has seen the changes in the weather and concluded that there is definitely something going on that science can explain. The mayor of Bastrop, Terry Orr, accepted the science that the climate is changing, though he was understandably wary of an issue that can be so politically divisive. Neither could recall it ever being discussed.

Cyndi Wright, the editor of the *Bastrop Advertiser*, was more doubtful, suspecting that the extreme weather was part of a natural cycle. She thought it was entirely inappropriate for discussion in her newspaper. "This is a community newspaper," she told me. "Sure, if climate change had a direct impact on us, we would definitely bring it in, but we are more centered around Bastrop County."

If climate change had a direct impact on us? This is surprising, that a journalist could not see any possible connection between the wildfire that had burned down her own house and an issue that scientists had, for twenty years, been warning would lead to increasing droughts and wildfires. Even Texas state climatologist John Nielsen-Gammon, who chose his words carefully, suggested the link and regarded the combination of extreme drought and record-breaking temperatures that fueled the fires as being "off the charts."

Of course no scientist will ever be able to say with total certainty that any single weather event is caused by climate change. But why does this prevent all discussion? What other topic is shut down because of a lack of total scientific certainty? Newspapers usually encourage debate, often ill-informed. Conversations are fueled by hunch and rumor. As I explore later, the lack of certainty is invariably an excuse for silence rather than the cause of it.

Nor were the people of Bastrop short of other things to say about the fire, including some highly conjectural opinions about who started it. Above all though, what they really wanted to share with me was their pride in their community and their capacity to overcome challenges. They spoke of the many acts of kindness, altruism, and generosity from strangers. Wendy Escobar told me how a customer at her cousin's hair salon in Longview sent her a thousand-dollar check in the mail. "The coolest thing to come out of the fire," she said, "was finding out how much people really cared, and how it's made people pull together so much."

One year later, Hurricane Sandy, the largest Atlantic hurricane on record, damaged or destroyed nearly 350,000 homes as it hit the New Jersey seashore. When I visited five months later, the destruction could still be seen everywhere in the small towns that line the shore.

In Seaside Heights the tangled remains of a roller coaster still lay out to sea, where it had fallen after the pier beneath it collapsed. Block after block of pastel-painted wooden houses were dark and abandoned, many homes twisted off their foundations or lying at crazy angles. Thirty miles north, in the small town of Highlands, the residents of the absurdly named Paradise Trailer Park had faced the full fury of the hurricane—one of them told me that he had survived the storm surge by sitting on top of his refrigerator. Now they had wrecked homes, no insurance payouts, and nowhere to go while the park's owner tried to evict them and redevelop the site.

In Sea Bright, just south along the coast, every shop on the main street was gutted, and the seawall was demolished. Two-thirds of the permanent residents were still homeless when I visited, and only eight of the hundred registered local businesses were back up and running.

Certainly, people were more inclined to accept that climate change exists in the Democrat territory of New Jersey than they were in Republican Texas. Dina Long, the charismatic mayor of Sea Bright, agreed that the frequency and power of the storms is changing and that the sea level is rising. Nonetheless, she could not recall anyone in her community discussing climate change in regards to the storm.

When I suggested to Long that she might band together with the leaders of other affected communities and demand federal action on climate change, she rolled her eyes. "Have you *seen* what Sandy did?" she demanded. "Climate change, *duh*, of *course* it is happening. But it is bigger than anything we could make a difference on. We just want to go home, and we will deal with all that lofty stuff some other day."

As in Bastrop, Texas, the dominant narrative all along the Jersey Shore was one of community cohesion and resilience. As Dina Long warmed to this theme, she flourished a small piece of plastic salvaged from Donovan's Reef, a landmark beachside bar. It is all that remains of the sign that used to hang over the door—a small fragment with two letters on it: "DO." Long brings out this talisman in every talk she makes to the townsfolk, media, and investors. It became the slogan of the Sea Bright Rising campaign and duly appears on T-shirts and posters around the town.

The strong sense of local pride I found in Bastrop and Sea Bright is entirely consistent with that found in other areas after disasters. Contrary to expectations, people rarely respond to natural disasters with panic, and there is often a marked fall in crime and other forms of antisocial behavior. People consistently tend to pull together, displaying unusual generosity and a sense of purpose.

These are times when people are most inclined to seek common ground and actively suppress the divisive and partisan issue of climate change. To talk about it seems inappropriate and exploitative, just as many people—President Obama's spokesman among them—refused to talk about gun control after the Sandy Hook school shooting.

The pain and loss of the event generates an intensified desire that there be a "normal" state to which one can return, making it even harder for people to accept that there are larger changes under way. The decision to stay, rebuild, and reinvest in that normality is accordingly validated by the community.

After losing all of his stock during Hurricane Sandy, Brian George, the owner of Northshore Menswear in Sea Bright, hung a sign outside his shop saying, "We love Sea Bright—we'll be back." After he reopened, business was great, he said, with many people buying something simply to thank him for staying. He accepted that climate change could bring more disasters but said he is resigned to it. "This is my home," he said, "and I guess we're just hoping another one doesn't come along any time soon."

Across the road, Frank Bain of Bain's Hardware also lost all of his stock—and found that his insurance did not cover floods. "I would have been better off if I'd burned the building down," he said bitterly. Bain, a much-loved pillar of his community, is a Republican and "no fan of Al Gore or his spotted owl," so he had always been unconvinced about climate change. After Hurricane Sandy, though, he had even stronger reasons for wanting to believe that it was just a rare extreme of nature: Not only had he rebuilt his store out of his savings, but he was "self-insuring"—putting money aside in the bank each year and hoping that the next storm was a long way off. He accepted that this was a gamble, but then again, being in business is a gamble, he said. "This is just how free enterprise works."

The extreme events themselves had already seemed like a gamble. In Bastrop and New Jersey alike, everyone was perplexed about how the wildfire or storm surge could destroy some houses and leave others untouched. "It was like Russian roulette," said Sharon Jones, sharing a birthday drink with her husband at the one bar still operating in Seaside Heights. Her house was entirely destroyed; the house across the road was left almost untouched. "Go figure," she said, raising her glass to the vagaries of fate.

After a disaster like Sandy or the Bastrop wildfires, people are presented with a stark choice about whether to admit defeat and leave or whether to stay and rebuild. When they decide to stay, as most people do, they are taking a gamble, and like any gamble, it predisposes them to undue optimism about the future and their own chances.

Psychological research finds that people who survive climate disasters, like people who escape car accidents unscathed, are prone to have a false sense of their own future invulnerability. A large field study in an Iowa town that had been hit by a Force 2 tornado found that most people had become convinced that they were less likely to be affected by a future tornado than people in other towns. The people in the areas that had suffered the most damage were often the most optimistic. So it is hardly surprising, following the extreme floods in 2012 in Queensland, Australia, that few people made any attempt to reduce their vulnerability to flooding, and many residents chose instead to spend their disaster relief and insurance premiums on general home improvements such as installing new kitchens.

Revealingly, then, extreme weather events provide an initial insight into why and how people can come to ignore climate change. At every stage their perceptions are shaped by their individual psychological coping mechanisms and the collective narratives that they shape with the people around them.

People yearn for normality and safety, and no one wants to be reminded of a growing global threat. As they rebuild their lives, they invest their hopes along with their savings in the belief that the catastrophe was a rare natural aberration.

At a community level they collectively choose to tell the positive stories of shared purpose and reconstruction and to suppress the divisive issue of climate change which would require them to question their values and way of life.

On reflection, it is hard to imagine any social environment in which a narrative of responsibility, austerity and future hardship would be less welcome than a community recovering from a climate disaster.

3

Speaking as a Layman

Why We Think That Extreme Weather Shows We Were Right All Along

"UNPRECEDENTED, UNTHINKABLE. THE DEVASTATION IS staggering. I struggle to find words." Choking back his tears, Yeb Saño, head of the Philippine government delegation, told the opening session of the November 2013 Warsaw Climate Change Conference of the devastation caused when Typhoon Haiyan hit his country three days earlier. He announced that he would fast in solidarity with the orphaned, the dead, and his own brother, who, he said, had still not eaten and had been gathering the bodies of the dead with his own hands. "To anyone who continues to deny the reality that is climate change, I dare you to get off your ivory tower and away from the comfort of your armchair and pay a visit to the Philippines right now."

Climate change can seem distant, uncertain, and incomprehensible. Saño made it seem real, immediate, and deeply moving. These personal stories and strong images, compounded by the constant repetition they received though the news media, spoke far more strongly to our sense of empathy and direct threat than the abstract data of graphs and scientific reports.

This is why climate change communicators are convinced that extreme weather events can, in the words of Elke Weber, an environmental risk specialist at Columbia University, "be counted on to be an

extremely effective teacher and motivator." Tony Leiserowitz, director of
the Yale Project on Climate Change Communication, calls them "teachable moments." Michael Brune, the executive director of the Sierra
Club, tells me that he sees severe weather as a kind of direct action.
"Obviously," he stresses, "not in an organized or manipulative way—
these are tragic events—but with the same capacity to change
consciousness and political direction."

Extreme weather events have already played a major role in the political momentum on climate change. In 1988, a severe drought and heat
wave across the Midwestern states provided the backdrop for Dr. James
Hansen of NASA to declare in a congressional hearing that there was a
99 percent certainty that global warming had already started. The rise of
consciousness about climate change in Europe that led to the signing of
the Framework Convention on Climate Change in 1992 was helped enor-
mously by severe storms in the spring of 1991, which were freely
interpreted by the media as a warning of the climate change to come.

Those campaigning for action on climate change do everything they
can to keep these connections alive in the public's minds. As mayor of
New York City, Michael Bloomberg personally approved the cover of the
November 1, 2012, *Bloomberg Businessweek*, with a picture of Hurricane
Sandy and bold block text reading, "IT'S GLOBAL WARMING, STUPID."
Referring to Hurricane Sandy, Al Gore said, "These storms—it's like a
nature hike through the Book of Revelation on the news every day now.
People are now connecting the dots."

Environment America global warming program director Nathan
Willcox is also convinced that "the more Americans see extreme weather
events in their backyards, the more they will reach out to their politicians
for action." However, his own research suggests that the relationship
between experience and conviction is far from straightforward. In the
seven years up until 2012, the Great Plains was consistently, and by a
wide margin, the region worst affected by climate-related disasters.
Nonetheless in the 2010 Senate elections, all the winning Republican
candidates for the Plains states publically refuted climate science or
opposed action to reduce greenhouse gas emissions.

Across the entire United States, the state most consistently affected by
extreme weather has been Oklahoma. In 2008, voters there were offered
a clear choice in their Senate election between Andrew Rice, a Democrat
candidate with a moderate but balanced acceptance of climate change,

and incumbent James Inhofe, the most active and aggressive climate denier* in the Senate. In a year when national concern about climate change was at an all-time high, Inhofe still won by a large margin, cleaning up in the five Oklahoma counties that were experiencing, on average, more than one federally declared weather emergency every year. As I write this, the so-called polar vortex is sweeping across the Midwest, and temperatures in Nowata, Oklahoma, have just fallen to 31 degrees below zero, three degrees lower than the previous state record. They keep getting hit and they keep voting for Inhofe.

Inhofe is as keen as any other campaigner to use climate events as a "teachable moment" for his own arguments. In February 2010, when an extreme blizzard deposited two feet of snow on Washington, D.C., Inhofe enjoyed some interactive family fun with his grandchildren by building an igloo on the National Mall. Alongside it, he erected a sign reading, "Al Gore's New Home!" and "Honk If You ♥ Global Warming."

At the same time, the West Coast was experiencing record-breaking warmth that forced organizers of the Winter Olympic Games in Vancouver to run fleets of trucks and helicopters, day and night, to bring in snow for the freestyle skiing and snowboarding events. Across North America as a whole there was enough evidence available to support any number of positions on climate change. Maybe *New York Times* columnist Thomas Friedman described the overall situation best when he simply called it "global weirding."

The problem is that, in a field normally dominated by technical specialists, weather events appear to be well within the range of laypeople's personal expertise. We might be in no position to judge the levels of trace greenhouse gases in the atmosphere, or sea levels, or the extent of glaciers, but we all think we know about the weather.

This is especially true in Britain, where, for some inexplicable reason, variations in our bland, damp weather are the subject of intense public interest. In his weekly opinion column in the *Daily Telegraph*, a national conservative newspaper, London mayor Boris Johnson pontificated on his own climate expertise:

---

* In this book I refer to those who, for ideological reasons, actively reject or undermine climate science, as *deniers*. I call those who legitimately raise scientific challenges, *skeptics*. And I recognize a third group of people who are simply not sure as the *unconvinced*. These are different kinds of people, with different motivations, and merit different titles.

Two days ago I was cycling through Trafalgar Square and saw icicles on the traffic lights; and though I am sure plenty of readers will say I am just unobservant, I don't think I have seen that before. Something appears to be up with our winter weather, and to call it "warming" is obviously to strain the language.

Johnson likes to cycle around town noticing things. When Franny Armstrong, director of the climate change documentary *The Age of Stupid*, was attacked by muggers, she was astonished to see the huge wild-haired bulk of the London mayor cycling into view, shouting, "Clear off, you oiks!" Johnson's public persona, you see, is that of a decent, up-for-a-laugh sort of fellow who makes up his own mind on the basis of common sense.

Johnson concluded his column, no doubt resonating like a tuning fork with the middle-age conservative readership of the *Telegraph*, by writing, "I am speaking only as a layman who wonders whether it might be time for government to start taking seriously the possibility—however remote—that the skeptics are right."

Because weather events can never be ascribed with certainty to climate change, we are therefore prone to interpret them in light of our prior assumptions and prejudices. If we regard climate change as a myth, we regard variable and extreme weather as proof that weather can be naturally variable and extreme. If we are disposed to accept that climate change is a real and growing threat, we are liable to regard extreme weather as evidence of a growing destabilization.

These selective processes are called *biases* by cognitive psychologists because they draw on preformed assumptions and intuitions to influence decisions. *Confirmation bias* is the tendency to actively "cherry-pick" the evidence that can support our existing knowledge, attitudes, and beliefs. These create a mental map—what psychologists would call a schema—and when we encounter new information we modify it to squeeze into this existing schema, a process psychologists call *biased assimilation*. We exercise both of these biases of selection and modification compulsively: to confirm our choice of restaurant, the attractiveness of our partner, the cleverness of our children, and to prove to ourselves that we have been "right all along" or that some personal mistake is "not as bad as all that." These two terms are subtly different in their academic usage, but, for ease of reading, I will use just one term for both: "confirmation bias."

Research finds that both of these cognitive biases are guiding our interpretation of extreme weather events and climate science as a whole. When asked about recent weather in their own area, people who are already disposed to believe in climate change will tend to say it's been warmer. People who are unconvinced about climate change will say it's been colder. Farmers in Illinois, invited to report their recent experiences of the weather, emphasized or played down extreme events depending on whether or not they accepted climate change.

Researchers discovered similar patterns in Britain. Interviews with flood victims in England found that their interpretation of the event largely depended on their views on climate change, and a wider poll found that Labour Party voters were twice as likely as Conservative voters to ascribe extreme weather to climate change. Consistent with my observations in Texas and New Jersey, people who had been personally affected were significantly less likely overall to ascribe it to climate change than those who were far away from the flooding.

Attitudes toward climate change are so politically polarized that it is not surprising that independents are the most likely to form views drawn from their direct experience of the weather. Sociologists at the University of New Hampshire found that 70 percent of independents were inclined to believe in human-caused climate change when they were asked about it on an unseasonably warm day. On abnormally cold days, that fell to 40 percent.

These contextual decisions display yet another form of bias—*availability bias*—that disposes people to make up their mind on the basis of the evidence that is most readily at hand. It can be just as misleading as any other form of confirmation bias, leading people to hugely overestimate the dangers of recent events and disregard the threat posed by more distant ones that they have not experienced.

Despite these biases, Tony Leiserowitz at Yale remains convinced that the teachable moment of changing weather is changing attitudes over the long term. He cites his own research showing that around two-thirds of Americans believe that global warming made specific extreme weather events worse. The highest number, not surprisingly, agree that the heat wave of 2011 and the warm winter of 2010–11 were linked to global warming.

These polls show that extreme weather is affecting ever-larger numbers of people and prompting them to consider climate change when the subject is raised. The larger question, though, is whether the growing experience of extreme weather will bring people together in a shared commitment to action, or whether their confirmation bias will push them even further apart. And if the weather extremes continue to intensify, whether the experience of coping with loss and anxiety will make people push it aside as something that they would rather not think about.

As the changes in the climate accelerate, new opportunities are emerging for us to engage or deny. Extreme weather events of entirely unprecedented scale and duration are now occurring regularly. Climatologists may be reluctant to ascribe a single weather event to climate change but are far more willing to agree that it is influencing widespread patterns of ever more extreme and bizarre weather.

As I complete this book, hail is falling in Cairo, snow in Israel, Syria, and Jordan. The United States is having the most extreme arctic cold it has ever experienced. Meanwhile Scandinavia has record-high temperatures, and Australia is entering its second year of unprecedented drought after temperatures reached so high that weather forecasters created a new color scale for the weather maps to accommodate them. Britain is ringed with more than a hundred flood alerts, and my hometown of Oxford has just had the wettest January since record keeping began in 1760. The day after I visited my nearby seaside town, the entire seafront was ripped apart by thirty-foot waves. The locals say they have never seen anything like it.

But they are still not talking about climate change. What *is* going on?

4

You Never Get to See the Whole Picture

How the Tea Party Fails to Notice the Greatest Threat to Its Values

I SPEND ALMOST ALL OF my working life with people who understand and accept climate change, so I decided to spend some time with people who are no less passionate in their conviction that we are completely wrong. This was how I came to find myself cruising along Texas State Highway 71, some thirty miles south from Bastrop, in the largest car I have ever seen: a seven-ton Ford Excursion, a car so huge that you need to lower a step before you can even climb inside.

My companions have little patience for environmentalists like me. We are arrogant, so arrogant, they said, to even think that we humans could possibly change this beautiful land enough to affect the world's weather systems. Our differences are directed by the selective vision of our respective confirmation bias—ironically the views to the left and right of our speeding SUV. Looking to the *right* they saw the wide-open fields and woodlands. Looking to the *left* I saw the railroad track that runs alongside the highway and a coal train reaching the end of its thousand-mile journey from Wyoming. The train was so long that I could see neither the front nor end, though I could see, silhouetted by the setting sun, the smoke stacks of its destination, the Fayette Power Project, pumping as much carbon dioxide into the air as the entire nation of Guatemala.

We were heading for the ranch home of Debra Medina, the feisty, fast-talking, take-no-prisoners mother of four who won one-fifth of the state vote as a wildcard candidate in the 2010 Texan gubernatorial election. On the first Friday of every month, forty Tea Party activists gather at her house to share home-baked food, their visions, and their frustrations—and to have a good laugh. It was with some trepidation that I accepted Debra's invitation to talk with them about climate change. I enjoy challenging audiences, but rarely ones that are this opinionated. Or this well armed. During her campaign for governor, Medina appeared across Texas TV channels waving her semiautomatic pistol, which is always loaded and ready to go. "It stays right here beside my car seat, where I can reach to it easily," she told the cameras, lifting the flap between the front seats where normal people keep their small change.

Scarcely two weeks before my visit, the Texan Republicans released their policy platform calling for protection from "Extreme Environmentalists," who purposefully disrupt the oil and gas industry, and demanding that climate change should be taught in schools only as "a challengeable scientific theory subject to change." This was going to be new territory for a former Greenpeace campaigner who founded a climate education charity.

So I presented Debra with a peace offering between our rival tribes: a King Edward VII tea caddy, and asked her to cough up two centuries in unpaid tax. Luckily, they laughed. Then I said, tell me what you think about climate change.

They hated *everything* about climate change: they hated the science, the scientists, Al Gore (especially Al Gore "and his garbage"), the United Nations, the government, solar power, the hypocritical environmentalists.

It was soon clear that climate change, or rather the narrative they had constructed around it, fit perfectly into a set of pre-existing ideological grievances about the distribution of power. The word they kept using was "control." James said that "carbon is a universal element that the government want to control." Denise added that climate change "is a convenient crisis. The government is using this as a tool of control." David said that the whole thing has been invented to create a "global tax for a one-world government"—this was clearly a familiar phrase and several people joined in to say it at the same time.

Which brought us rapidly to Agenda 21, a long, dull, and largely forgotten resolution proposing local goals for sustainable development that emerged from the 1992 U.N. Conference on Environment and

Development. To the Tea Partiers, Agenda 21 is the constitution for a one-world government containing the detailed plan for how "they" will create the issue of climate change to control us and suppress our local freedoms. It is unthinkable to them that there would not be a constitution of some kind for global dominance—after all they regard the U.S. Constitution as a sacred text and can quote it from memory. After the meeting, Dave signed his own copy and presented it to me. It was exactly the same color and shape as my British passport, causing predictable confusion later on in my travels.

But even with a written constitution to hand, the truth remains complex and elusive, because, they tell me: "You never see the whole picture—you have to draw the line a little segment at a time." They maintain that they have to be constantly vigilant and ask questions: "We're not anti-intellectual people in this group. We want to know the truth. We think outside the box and search for our own answers."

And they certainly ask a *lot* of questions—it is a hallmark of their conversational style. People's statements frequently broke down into a string of questions: Which way is the wind blowing? Where does the money come from? What happened to the scientists? What happened to their opinion? Could they also have been misled? Or they could be mistaken? What's the baseline? What is normal? When was normal? Was there ever supposed to be a normal?

This admirable willingness to challenge things makes them feel somewhat superior to other people. They said that the reason people believe in global warming is that they aren't logically minded and are "just not educated properly at school anymore."

Like climate scientists, or environmentalists, these Tea Partiers stress the overwhelming importance of information. The problem is that it is so hard to get the right information—meaning they have to get it from people who share their values: "My favorite radio show host, Dave Champion, always says, 'The government lies. It lies all the time, and it lies even when the truth would serve it better.'" So, all the conventional providers of information are corrupt and suspect, and, they say, scientists know all too well that "if you can get the population scared to death, they will be willing to write a check for their research."

Passion is a word they use frequently: "The passion is not that we cover our ears with our hands and don't want to hear the facts. The passion is we don't want to be controlled." They are especially passionate about their independence. One man said, "I'm not with the environmentalists. I'm not with the oil companies. I did not come to take sides; I came to *take over!*" Everyone loved this, and the whole room erupted into laughter, claps, and cheers.

It is easy to focus on differences, and certainly rural Texan Tea Partiers are quite unlike urban liberal environmentalists. But the real surprise for me was to discover that being with them felt entirely familiar. They have exactly the same boisterous, opinionated, autodidactic, and tribal spirit as the grassroots environmental campaigners I have worked with in campaigns to save forests, stop open-pit coal mining, block new superstores, and, yes, demand action on climate change.

And they have plenty in common with environmental activists in their political instincts. They are outsiders driven by their values, defensive of their rights, and deeply distrustful of government and corporations: ExxonMobil and Monsanto both came up for attack in our conversation. Indeed, strange alliances had already been built around the campaign against the Keystone XL pipeline, which is opposed by environmentalists for its contribution to the carbon economy and by the Texan Tea Partiers for its use of eminent domain to seize land from property owners.

While the Tea Partiers had lots of questions, I left with just one of my own: What had led them to reject the one issue that, above all others, truly threatens the things that are most important to them: family, property, freedom, their beloved country, and God's creation—one, what is more, that has reached this critical stage because of the thing they hate the most: government and corporate self-interest?

Is it because climate change feels too far away? Perhaps, though the Tea Partiers are quite prepared to agitate about other complex international issues that catch their fancy. Is it because they feel powerless to do anything about it? Probably not, as they seem to thrive under conditions of powerlessness. Is it because it is depressing and frightening? Hardly: The entire Tea Party movement is built on fear and the warnings of disaster.

Is it because it is scientific and technical? No, these are people who willingly seek out information. Is it because climate change is contested and uncertain? Absolutely not: To be honest, the Tea Partiers appear to be entirely capable of believing any number of uncertain things on very limited evidence.

The answer must lie elsewhere—not with the issue itself but with the way it has been told. It must be something about the way the story of climate change has been constructed and communicated, the people who tell it, and how it has attached itself to their values.

5

Polluting the Message

How Science Becomes Infected with Social Meaning

PROFESSOR DAN KAHAN, THE LEADING light of the Yale Cultural Cognition Project, is an expert on how conflicting cultural values influence decision making. So when I notice that a garish plastic figurine of Gene Simmons from the 1980s übercamp rock band Kiss has taken pride of place on the mantelpiece of his sober Yale Law School study, I ask him if this is some ironic academic joke—maybe a comment on the disjuncture between Simmons's support for George Bush and his anti-authoritarian stage persona? "Nah," Kahan says, "I just like him . . . because he *rocks*!"

This is how Kahan speaks—at very high speed, in a hyperintelligent soup of cognitive jargon and hip slang. Clearly he is not someone who fears challenging conventions or crossing cultural boundaries.

For Kahan, the defining quality of climate change is not any lack of overall concern—he says there is plenty of that. Nor does he agree with the opinion of many activists that the key influence on attitudes is the politicization of the media coverage. "Face it," he says, "even if it does get mentioned on MSNBC or Fox News, ten times more people will always be watching funny animals."

Kahan is a cultural omnivore and is intrigued by funny animal videos. He urges me to watch "The Crazy Nastyass Honey Badger" on YouTube because "it's even more bad-ass than Gene Simmons." More than sixty-five million people have watched that honey badger video. Over on the Intergovernmental Panel on Climate Change channel, the climate scientists have a hard time reaching an audience numbering in the four digits with their decidedly un-crazy-ass slideshows.

For Kahan, the reason why people do not accept climate change is nothing to do with the information—it is the cultural coding that it contains. He argues that people obtain their information through the people they trust, or, beyond that, from the parts of the wider media that speak to their worldview and values. Most of the time, this is a highly effective shortcut and works fine, unless, in Kahan's words, the information becomes "contaminated" with additional social meaning and becomes a marker of group identity.

Kahan cites gun control as a case in point. Polls in West Virginia show that 65 percent of people want more gun control but, he says, you would be a fool to run for election in that state campaigning for gun control. "What you don't know—and no poll has told you—is that *85 percent* of people in West Virginia know that you can't *trust* politicians who say that they want gun control."

Attitudes on climate change, he argues, have become a social cue like gun control: a shorthand for figuring out who is in our group and cares about us. Just because polling shows a high level of concern about the issue does not mean that there is an equally high level of support for the people who promote it.

Kahan's extensive work on understanding people's resistance to vaccination forms a direct analogue for how they form their opinion on climate change. There are few issues in which the science has become so contaminated so rapidly. In Britain a single research paper in 1998 arguing that the combined mumps, measles, and rubella (MMR) vaccine might cause autism in children was accepted as proof by one-quarter of the public, and immunization rates plummeted. Scientific data was soon abandoned in the dirty public battle that contrasted the cold, mechanistic approach of the scientists with the raw emotional appeal of the parents convinced that their children had changed immediately after their immunization shot. Fifty percent of people took the presence of a media-generated debate as evidence that the science was in doubt.

In the United States, there was a similar disaster when the state of Virginia decided that the package of compulsory vaccinations for entry into middle school should include one against human papilloma virus, a very common sexually transmitted disease that causes cervical cancer.

So, you have the government knocking at the door of a conservative Christian community, saying, according to Kahan, "You know your twelve-year-old daughter? Well, she's going to be having *sex* in the next year and getting a venereal disease, so we're going to give her a shot. And if you don't like that, she can't come to school." This was a toxic brew of government interference, moral challenge, and offensiveness.

The lessons for climate change are clear. First, rational scientific data can lose against a compelling emotional story that speaks to people's core values. As I discuss later in the book, these cultural meanings become deeply attached and therefore cannot be removed by applying more scientific argument.

Second, communications from people's family, friends, and those they regard as being like themselves (their peers) can have far more influence on their views than the warnings of experts'.

Third, attitudes toward climate change fit into a larger matrix of values, politics, and lifestyles. Thus, as Kahan, Leiserowitz, and others at Yale argue, there are identifiable "interpretive communities": people who believe or disbelieve in climate change—and one can predict with some accuracy who they are, how they live, who they trust, and where they receive their information. Over the past ten years, detailed profiles have emerged.

Homo credens (the convinced) are most likely to be middle-age, college-educated liberal Democrats. Women are more likely to be believers, which is consistent with the observation that women tend to be responsive to other health, safety, financial, and ethical risks.

Homo negator (the unconvinced) are almost always strongly conservative in politics—very few are not—and tend to be from the more affluent and powerful social groups. They are very likely to be men and may display a low level of risk perception in other areas. This is a familiar group to risk researchers, who have named the "white man effect" after the danger that men in this group can seriously distort their social research.

Putting it together, one could predict that middle-age male motorbike riders are not well disposed to believe in climate change even before reading a Canadian survey that found that, indeed, two-thirds of them did not accept climate change.

Many other studies have identified further attitudinal subgroups (one study names them the Cautious, the Doubtful, the Alarmed, and the

Disengaged) each with their own sociopolitical demographic and distinct values.

The fact that attitudes to climate change can be predicted by such specific cultural characteristics is further evidence for Kahan's argument that the science has become polluted with social meaning. Understanding how attitudes to climate change are acquired and held—and how they might be changed—therefore requires understanding how people's social identity comes to have such an extraordinary hold over their behaviors and views.

6

The Jury of Our Peers

How We Follow the People Around Us

IN THE EARLY HOURS OF the morning of March 13, 1964, Kitty Genovese was assaulted and then stabbed repeatedly in a densely populated residential area of Queens, New York. Thirty-eight people (one of them ironically named Joseph Fink) said they had heard her screams and done nothing to intervene. One man lamely shouted, "Let that girl alone," out of his window before going back to bed. Another pulled a chair up to the window and turned out the light to better see what was happening. No one thought to call the police until it was too late.

Rather than being a sad testament of a broken society—as the newspapers of the day suggested—this lack of response actually revealed the strength of social conformity. People read the social cues. They saw that no one else was taking any action and decided that it was in their best interest to keep out of a potentially dangerous situation. Knowing that others had heard the cries, they diffused responsibility, assuming, quite wrongly, as it turned out, that someone else had called the police.

The tragic Genovese incident launched a rich and still expanding body of research into the importance of social cues in defining what issues people respond to and what ones they ignore. It is a fascinating feature of this *bystander effect*—as it was subsequently named—that the more people we assume know about a problem, the more likely we are to ignore our own judgment and watch the behavior of others to identify an appropriate response.

A string of experiments confirmed the power of the bystander effect. In one particularly entertaining experiment, an actor faked having a seizure over the laboratory intercom. The last words heard from him were "I could really—er—use some help, so if somebody would—er— give me a little h-help uh er er . . . I'm gonna die," followed by a choking noise and silence. Of fifteen participants in the experiment, six never got out of their booths, and five others only came out well after the "seizure victim" apparently choked.

Of course, you can only run these kinds of experiments for a few years before your subjects start to get wise to the trick, especially if they are psychology students. Years later, when a subject in a psychology experiment had a real epileptic fit, the other participants were convinced that it was being faked for the experiment and refused to get off their chairs.

Climate change is a global problem that requires a collective response and so is especially prone to this bystander effect. When we become aware of the issue, we scan the people around us for social cues to guide our own response: looking for evidence of what they do, what they say, and, conversely, what they do *not* do and do *not* say. These cues can also be codified into rules that define the behaviors that are expected or are inappropriate—the social norm. If we see that other people are alarmed or taking action, we may follow them. If they are indifferent or inactive, we will follow that cue too.

This social conformity is not some preference or choice. This is a strong behavioral instinct that is built into our core psychology, and most of the time we are not even aware that it is operating. It originated as a defense mechanism during our evolutionary development, when our survival depended entirely on the protection and security of our social group. Under such conditions, being out of sync with the people around us carried a potentially life-threatening danger of ostracism or abandonment.

There are, therefore, real and serious risks involved with holding views that are out of step with your social group and your brain is wired to give them greater weight than other risks, even those that directly threaten you. In experiments on social conformity, people chose to adhere to a social norm even under conditions when there was a real and imminent external threat, such as smoke coming from under the door.

So if your views on climate change differ from the socially held views, you find yourself balancing two risks: the uncertain and diffused risk of climate change as opposed to the certain and very personal social risk of opposing the norm. As I will show, people often decide that it may be better to say nothing at all about climate change, even with their close friends.

Although conformity is important for functioning societies, a small number of dissenters are required to identify new threats. In the famous Hans Christian Andersen story "The Emperor's New Clothes," a small boy has been given the social license to declare that the emperor is naked. Andersen based his story on a Spanish folk tale in which a Moor (an African Muslim) was permitted, by virtue of his outsider status, to defy the social norm. In our own times, nonprofit organizations, such as environmental and human rights organizations, are given some license, even in repressive societies, to raise challenging questions, providing that they remain peripheral.

Andersen added his own astute variation to the original Spanish tale. After the boy shouted about the Emperor's nakedness, "the Emperor was vexed, for he knew that the people were right; but he thought the procession must go on now! And the lords of the bedchamber took greater pains than ever, to appear holding up a train, although, in reality, there was no train to hold."

The final moral of "The Emperor's New Clothes," then, is that these social norms are highly resilient to change—even when the norm has been effectively challenged, the social cost of admitting a mistake and the effort required to change a behavior may be so great that it is easier to continue with a known lie.

One way of ensuring against such a challenge is to surround yourself with people who agree with you. In our dispersed and media-driven society, people are able to immerse themselves in a self-constructed social network where the norm is entirely consistent with their own views. They restrict their information sources to carefully selected news media, websites, blogs, and publications—the so-called echo chamber—that reinforce their views. Tea Party members and environmental activists alike share a distrust of the mainstream media and depend on information sources that speak specifically to their interests and values.

Researchers in Australia found that these self-constructed networks had created what they call a "false consensus" effect around climate change, which led both sides to believe that their opinion was more common than it actually was. However, because the loud and very vocal climate change deniers were also heard far into the mainstream media, both sides tended to hugely overestimate their numbers, guessing them to make up a quarter of the population. In fact they made up less than 7 percent.

When people misread the social norm in this way, it can lead them to suppress their own views, thus widening the divide and further reinforcing the false consensus—and at its most extreme, creating a society in which the majority of people keep silent because they fear that they are in the minority. This process, known as *pluralistic ignorance*, helps to explain the extreme polarization around key markers of political identity such as abortion, gun control, and, increasingly, climate change.

Communicators have long hoped to harness the power of social norms and conformity to steer people away from high-carbon behaviors. They argue that this is particularly appropriate for collective issues, like climate change, in which people require proof that others are contributing before acting—called *conditional cooperation* in the literature. In a widely influential experiment, Robert Cialdini, professor of behavioral psychology at Arizona State University, placed hangers bearing different messages on the towel racks in motel rooms asking people to reuse their towels. By far the most successful message was the one that appealed to a social norm with the message that 75 percent of guests "help save the environment" by reusing their towels. Even then, less than half of them did so, suggesting that people require more evidence of a norm than can be provided by a single hanger tag.

In 2010, the consultancy Opower built on Cialdini's experiment to use reported social norms to encourage energy conservation. It persuaded Connexus Energy, a Minnesota-based utility, to include with its customers' electricity bills a report on how their energy consumption compared with that of their hundred nearest neighbors. To prevent backsliding, people who had lower than average consumption received reports covered in smiley faces and with the exhortation "Great." After all, Opower reckoned, who would ever want to disappoint those smiley faces?

But manipulating norms in this way also has costs. The tactic (which achieved only a paltry 2 percent energy savings) made no attempt to strengthen shared values. It is not surprising that, confronted with all that green messaging and those smiley faces, some conservatives *increased* their energy consumption—apparently as an act of defiance.

It should already be clear that social norms might be powerful, but that people are correspondingly extremely alert to the cultural codes that they carry. This is why drawing too much attention to an undesirable norm can seriously backfire. When park rangers erected a sign in Arizona's Petrified Forest National Park that read, "Your heritage is being vandalized every day by theft losses of petrified wood of 14 tons a year, mostly a small piece at a time," the rate of theft significantly increased. Although the sign attempted to communicate the undesirability of theft, what it actually communicated far more powerfully was that stealing a small amount of wood was a perfectly normal activity.

Environmental organizations never seem to learn this message. In 2007 the Alliance for Climate Protection, founded by Al Gore, ran a commercial in which young parents at a smart dinner party list the reasons why climate change is a myth while tossing their leftovers onto the heads of their children sitting behind them. The final tagline was "What kind of mess are we leaving our children?" The ad was meant to be ironic but was actually a spectacular mistake—a thirty-second promotion for the wrong arguments that presented climate deniers as attractive young suburban professionals.

Maybe with this in mind, the Alliance's next foray into social norm campaigning was based around common values and appeals to national unity. The three-year advertising program, supported by a staggering three-hundred-million-dollar budget, aimed to recruit ten million advocates for national climate change legislation. It was called We Can Solve It.

The campaign's name was a combination of Barack Obama's campaign slogan "Yes we can" and the Second World War slogan "We can do it!"—forever associated with the iconic poster of the bicep-flexing Rosie the Riveter. Its advertisements drew on other familiar historical images of collective purpose, such as the Normandy landings, civil rights marches, and the Apollo landing, and showed bitter political rivals such as Nancy Pelosi and Newt Gingrich, or Al Sharpton and Pat Robertson, smiling on a couch together and agreeing to cooperate.

It was a brief cease-fire before the growing partisan divide on climate change forced Gingrich and Robertson to issue rebuttals. But it was, at least, a bold attempt to create a social norm that climate change is a historic challenge that can be overcome by American innovation and the "can-do" spirit.

However, the repeated use of the pronoun *we* was more problematic. I call this the "slippery *we*" because it sounds inclusive and assertive when read in a transcript, but it is actually ambiguous and often meaningless. Unlike the "conventional *we*," which describes a common action or attitude (in which sense it appears throughout this book), the "slippery *we*" is a rhetorical gambit to create a sense of norm while demonstrating what management manuals like to call *transformational leadership*.

The political use of *we* has been escalating with a kind of rhetorical desperation. In their inauguration speeches, George Washington said *we* once only, Jefferson and Lincoln around ten times, John F. Kennedy twenty-nine times, Barack Obama fifty-seven times.

In his keynote policy statement on climate change, delivered at Georgetown University in June 2013, Obama went on an unparalleled *we* spree, using it ninety-six times, sometimes in first-person pileups: "We can figure this out. We've got to look after our children; we have to look after our future; and we have to grow the economy and create jobs. We can do all of that as long as we don't fear the future; instead we seize it."

But who is Obama's *we*? Is it him and his administration, his supporters, the American people, or humanity as a whole? Without such clarity, this is the language of the bystander looking out the window and saying, "We really must do something about this."

The slippery *we* can be deeply alienating for the people who do not consider themselves to be included within it. Unlike the languages of truly cooperative cultures, such as indigenous Australians and Native American societies, English has no means to differentiate between the inclusive *we* (me and you and your group) and the exclusive *we* (me and my group but *not* you).

If you oppose Obama, his avowal of common purpose sounds deeply exclusive and you hear him saying, *me and my fellow global warming zealot cronies* are going to force *you* to do this. No doubt if you are a supporter of Obama, it sounds wonderfully inclusive. And if you are, consider how you feel about this equally stirring rhetoric: "We can and must work together, and re-chart our course toward a better future. We will begin to thrive again when we begin to believe in ourselves again." When I tell you that it is from a 2013 speech by Tea Party founder and climate denier Rand Paul, do you feel embraced in his "we" or repelled?

While politicians use the slippery *we* to create a false norm of action, many people use it to create a false social norm of inaction. For example, a woman in a Swiss focus group explained why action on climate change is pointless with the following words: "We just consume. We are somehow helpless. We don't care anyway, as we don't exactly know what effects we cause. If we took every problem equally seriously, we would become permanently depressed." She is quite deliberately playing with the power of the social norm, projecting her own views onto a supposed "we" and being resigned in her inability to challenge the norm she has just fabricated.

In focus groups, people who stumble to explain their own inactivity suddenly become fluent in explaining the reasons why *other* people are unable to act, mobilizing the language of popular psychology and talking about anxiety and denial. In a 2012 study one woman, described as being well educated and middle-class, said, "So yeah, I don't think people take it seriously because it's not a thing that affects you here and now, and I think people often react slowly or badly to things that seem very distant."

There are two layers of distancing here. There is the legitimate observation that climate change as an issue doesn't *feel* dangerous—an issue I address in the next section of the book. But there is also her own detachment that enables her to make an observation about why *other people* do not react. Conveniently, this reading of the social norm helps to justify her own inaction.

So, a few words of warning. Although our instincts lead us to seek out social cues when forming our views on climate change, these could be deeply misleading. The jury of our peers is hardly impartial, and through our confirmation bias, we could very well be choosing to read the norm in the way that best suits the position that we have already decided to hold.

7

The Power of the Mob

How Bullies Hide in the Crowd

PEOPLE NOT ONLY IDENTIFY STRONGLY with their own social group but also believe that it has a distinctive identity that makes it superior to other groups. *Self-categorization theory* recognizes that there are two processes at work. First, we seek to achieve closeness and similarity with people with whom we feel an identity and kinship: our in-group. Then we seek to establish our differences from the people who are not like us: the out-groups. Our attitudes and behaviors are shaped by the people around us who we want to be like as well as by the people beyond us who we want to be *unlike*.

The effect of this on environmental attitudes was shown in a clever experiment in Britain. When participants compared themselves with people in Sweden—who are generally considered to be highly environmentally aware—they showed *less* interest in energy conservation. In contrast, when they compared themselves with Americans, who they regarded as energy wasters (sorry about this—we are talking here about cultural stereotypes), they suddenly found a zeal for all things green. In other words, people in the in-group sought to move in the opposite direction to the out-group. As well as keeping up with the Joneses, they wanted to keep well away from those clean living Olafsons and those gas-guzzling Yanks.

This in-group and out-group behavior is apparent in all attitudes to the climate change issue. It leads both sides to underestimate the diversity of views within both their own ranks and those of their opponents, creating false stereotypes around liberal environmentalists and conservative deniers. And it leads them to exaggerate their own worthiness and denigrate their opponents.

The Internet has produced entirely new areas for the formation, expression, and enforcement of social norms and in-group, out-group dynamics. Facebook enables people to broadcast their views far more widely and brazenly than would happen in typical social interactions.

An aggressively contested social norm is at work in the swarms of comments that follow every article on climate change. Experiments have shown that the insertion of aggressive comments into descriptions of controversial issues does nothing to change people's views but greatly increases their in-group identification with the view they already hold.

When scientists post a research paper on the Internet about climate change deniers, the angry responses generate even more data about climate change deniers—like a fast breeder reactor. The psychologist Stephan Lewandowsky received enough aggressive responses to his first research paper on climate denial conspiracy theories to provide the basis for yet another study. It bears the enticing title "Conspiracist Ideation in the Blogosphere in Response to Research on Conspiracist Ideation."

While the bystander effect emerges from a sense of shared powerlessness, and a sense of shared power enables a range of abuses and violence, the anonymity of the new electronic norm enables outright bullying—abusive and violent e-mails received by high-profile scientists and activists calling them "Nazi climate murderers" and telling them to "go gargle razor blades."

The late Stephen Schneider, one of the highest-profile climate scientists in the United States, found his name on a "death list" with other Jewish climate scientists on a neo-Nazi website. He had his address unlisted and had extra alarms put on his house. "What else could I do?" he asked. "Wear a bulletproof jacket? Learn to shoot a Magnum?"

Things came to a head around unfounded allegations in late 2009 that climate scientists had been distorting data. Glenn Beck on Fox TV called on scientists to commit suicide; the late Andrew Breitbart, a right-wing provocateur, tweeted, "Capital punishment for Dr. James Hansen"; and the blogger Marc Morano called for climate scientists to be publically flogged.

Bill McKibben, America's most prominent climate change activist and the founder of 350.org, is characteristically phlegmatic: "I think my working theory is that if someone really wanted to shoot you, they probably wouldn't send you an e-mail first." I am not sure how much comfort one can find in that thought.

Something is at work here that is far more powerful, and more toxic, than the usual antagonism between different groups. Scientists are not an oppositional culture, and they work exceptionally hard to stay outside political divisions. They are, according to every opinion poll, by far the most respected and trusted of all the professions. The way that climate scientists have been treated is exceptional, and unparalleled in the recent history of science. Louis Pasteur never considered learning how to use firearms; Jonas Salk did not need to fortify his house. Scientists are not enemies and have never sought to be. They have been set up to play that role in a climate story line that, it would seem, cannot refute climate change without demonizing the people who warn us about it.

8

Through a Glass Darkly

The Strange Mirror World of Climate Deniers

FEW PEOPLE HAVE BEEN MORE influential in writing the story of climate change as a struggle between good and evil than Myron Ebell, the director of global warming at the libertarian think tank The Competitive Enterprise Institute (CEI). Ebell is up for the fight, and his biography on the CEI website happily includes a string of accolades from his critics. Greenpeace calls him a "climate criminal," *Rolling Stone* calls him "a leading misleader," and *Business Insider* anoints him as "enemy #1 to the climate change community."

Meeting him, I am struck once again by how closely the opponents of action on climate change mirror the advocates. Just as the Texan Tea Partiers reminded me of other grassroots campaigners, Ebell reminds me of other smart-suited policy wonks in the environmental organizations he despises—all of them fighting over the same narrative battleground with their reports, press releases, strategic lawsuits, conferences, slogans, and videos.

While waiting for our meeting, I get a glance inside Ebell's office—a very brief peek, as he moves me on rapidly. It is possibly the messiest office I have ever seen, looking as if a hurricane (most assuredly *not* caused by climate change) had blasted through an entire floor and dropped everything in one room. Ebell is meticulously turned out in a smart suit, but his office speaks of pure campaigner.

Ebell is keen to stress that his principles are pure and uncorrupted by vested interests. Yes, he freely admits, he has received funding from ExxonMobil for his climate change work, but he was pursuing the same issues before it funded him and after it stopped funding him. Environmentalists, on the other hand, are hypocrites who denounce fossil fuels while taking money from oil and gas companies when they choose to.

He cites the story that the Baptists and the bootleggers worked together out of mutual interest to demand alcohol prohibition laws. It is a myth (deliberately concocted by the free market economist Bruce Yandle) that is much used by libertarian organizations because it suggests that anyone demanding regulation is morally corrupt.

Ebell is principally concerned with describing climate change as a battle of political principles. "The environment movement," Ebell tells me, "is not an objective, well-intentioned movement that cares about saving the planet." It emerges from the New Left, and regardless of the issue, it always proposes the same solutions: more government control, more power for the technocratic elite, and less material standards of living for people. He and his colleagues are, he says, involved in "a David versus Goliath struggle" against big government and corrupt environmentalism.

David and Goliath? I must look a bit taken aback to hear him use the favorite biblical metaphor of progressive social rights organizations. So Ebell repeats himself—yes, he says, David and Goliath.

What is more, notes Ebell, who is now on a roll, his side plays a "decent game." A decent game, apparently, includes persistent personal attacks on the integrity of climate scientists. The latest tactic, championed by his CEI colleague Christopher Horner, is bombarding high-profile climate scientists with aggressive lawsuits demanding access to their private correspondence.

Ebell is convinced of his virtue. He insists his side is just being critical but would never stoop so low as to smear individuals—unlike environmentalists. And as he says this, he produces with a flourish a printout of a blog I had written that is moderately rude about him and his colleagues. He then takes great pleasure in reading it back to me line by line. Now I see why Ebell agreed to this interview.

Our discussion is marked by a banter in which every criticism that might be made by climate change campaigners is repeated and returned with interest. Greens are corrupt. Greens are political extremists. Greens distort the science for their own ends. Skeptics—because Ebell certainly would not consider himself a denier—are the underdog in a corrupt world fighting for a just cause.

These stories are dominated by enemies—the titanic struggle of good versus evil. The issue comes alive through the battle, and the characters and information, it would seem, attach themselves as convenient background for the story line that already exists.

For conservatives, climate change appeared as an issue at just the right time to replace the Red Menace bogeyman that had so long been their mobilizing enemy. Scarcely one year after the collapse of the Soviet Union, the 1992 Earth Summit provided a replacement threat—a shift of cast members in the longstanding opera of international ideological menace. As Rush Limbaugh says, climate science "has become a home for displaced socialists and communists."

Those on the far right have a particular attachment to demonizing their political opponents. In 2011 the website Right Wing News surveyed forty-three popular conservative bloggers to determine the "worst figures in American history." Jimmy Carter, Barack Obama, and Franklin D. Roosevelt led the tally, all well ahead of Benedict Arnold, Timothy McVeigh, and John Wilkes Booth.

The inability to differentiate between psychopathic killers and ideological opponents is perfectly exemplified in a notorious billboard erected by the Heartland Institute in Chicago with the photograph of the murderer Ted Kaczynski (the Unabomber) and the caption "I still believe in global warming, do you?" This ludicrous advertisement has inspired numerous Internet parodies, such as a photo of Adolf Hitler with the quote "I still believe kittens are cute, do you?" So much for the decent game.

As our meeting closes, Ebell offers me a bowl containing four small round chocolates. The foil around them is printed with a map of the world to make them look like little globes. I look at them sitting there and recall all those infographics in green publications of the four planets' worth of resources that we consume each year. He smiles at me—is this a test, I wonder, an initiation rite? Then again, a chocolate is still a chocolate, so I scoff the lot. After all, there is only so much metaphor that a man can bear.

9

Inside the Elephant

Why We Keep Searching for Enemies

CLIMATE CHANGE—THE REAL CLIMATE change based on scientific facts—lacks any readily identifiable external enemy or motive and has dispersed responsibility and diffused impacts. Issues of this kind are notoriously hard to motivate and mobilize people around. For example, one of the biggest killers in the world is the smoke from indoor cooking stoves. It kills 1.6 million people every year. But it has no enemy, no one is to blame, no one has responsibility—and very little is done to prevent it.

The lack of a clear enemy poses a problem for the news media trying to report on climate change. Mark Brayne, a former senior BBC correspondent, explains that journalism needs events, clear causes, and "a narrative of baddies and goodies." However, climate change has none of these. "It is slow moving, complex, and what's more, we ourselves are the baddies. That's not something listeners and viewers want or wanted to be told," Brayne says.

In a polarized battle, each side constantly measures itself against its opponent, learning from each other and then adopting the same narratives. This pattern of mirroring, sometimes called inversionism, is familiar from other polarized debates such anti-smoking, gun control, and abortion. Reading through my own transcripts and dozens of other interviews, I can identify a template that could be equally used by either side when it talks about the other side:

They (the other side) needed a new enemy after the end of the cold war and needed a political cause that would enable them to exercise political influence. So they created a story around their political worldview designed to play to people's fears and weaknesses with us as the enemy. They try to play the moral high ground but their real motives are money and political influence. They claim that they are weak, but actually they are much more powerful than us because they have the support of large funders with overt political interests and because they are promoted by a lazy and biased media. We get abused and sometimes even get hate mail and death threats, but it's our duty to expose these lies in the interest of the world's poorest people and to save civilization from the greatest threat it has ever faced.

And everyone defends the science—or rather their own science. The language used by Rex Tillerson, CEO of ExxonMobil, could be equally well adopted by any number of climate scientists or activists. The public, he complains, is illiterate in the areas of science, math, and engineering. Interested parties take advantage of this ignorance to "manufacture fear" supported by "a lazy and unhelpful media who are unwilling to do the homework."

Scientists and mainstream environmentalists share this belief that pure and accurate information is the wellspring of public attitudes and government policy; they regard those who pollute that information or prevent its flow as the enemy. When I asked James Hansen, then a NASA climate scientist, why people did not yet accept climate change, he said, "The answer is very simple—it's money. The fossil fuel industry is making so much money that they control our governments, the media, and everything they tell us."

Few scientists, not even Hansen, have been so publicly abused as Michael Mann, the director of the Earth System Science Center at Pennsylvania State University, whose iconic "hockey stick chart" mapped temperature changes over the past thousand years. He has been pilloried on Fox News and in the Senate, portrayed as a dancing puppet on YouTube videos, and received, by his own reckoning, thousands of abusive e-mails, including demands that he commit suicide or be "shot, quartered and fed to the pigs, along with your family." When I invited him to find a metaphor for this struggle, he settled, without any hesitation, on *The Lord of the Rings*. "It's a classic tale of the struggle between good and evil, but the stakes are the earth itself. The CEOs of fossil fuel companies who fund a disinformation campaign to confuse the public are the forces of Mordor. The scientists are Gandalf." I would add that, with his goatee beard and twinkling eyes, Mann could also find a place in that battle—a climatological faun, maybe.

This disinformation campaign, often referred to as the "denial machine," contains a wide network of think tanks such as CEI, media outlets, and politicians. But of late, campaigners have consistently set their sights on its most prominent and nefarious funders: David and Charles Koch, sibling inheritors of the second largest privately owned company in the United States.

The Kochs like to spend a small amount of their eighty-billion-dollar wealth on their favored political causes, including the Tea Party, political action committee advertising, and the libertarian think tanks opposing action on climate change, into which they have poured some sixty-seven million dollars since 1997. Not surprisingly, the Kochs are the number-one hate figures of the progressive left and environmentalists alike, and the grinning brothers are often portrayed in activist literature as the twin heads of the "Kochtopus," surrounded by the spreading tentacles of their gas, oil, and chemical interests. This is the latest in a long cartoon history of rampaging corporate cephalopods, which have included railroad monopolies, ice monopolies, Tammany Hall crooks, Standard Oil, and—campaigners would be horrified to realize—the international Jewish "conspiracy."

Certainly I would never claim that the Kochs are not major political operators or that the "denial machine" they help fund has not played a significant role in shaping public opinion, as has been superbly documented in recent books such as *The Merchants of Doubt, Heads in the Sand*, and *Climate Cover-Up*. And oil companies *are* blocking action. There *is* a well-funded politically motivated campaign that distorts and pollutes the science.

But I would argue that the constant goading of the aggressive deniers over the past twenty years has led campaigners and scientists alike to invest too much of their emotional energy into this single struggle and to forget that there is an infinite number of different stories that have yet to be told, and that the vast majority of people are being entirely ignored during their punch-up.

All of these enemy narratives seem entirely natural to the people who hold them. They merge seamlessly with their existing values and belief systems, build on the metaphors of previous struggles and their own reading of history. I share many of these values and this sense of history. I have spent much of my working life in the environmental movement, leading campaigns against governments, corporations, and international finance. Many issues do come down, in the end, to a struggle against distinct and identifiable vested interests.

But climate change *is* different. The missing truth, deliberately avoided in these enemy narratives, is that in high-carbon societies, everyone contributes to the emissions that cause the problem and everyone has a strong reason to ignore the problem or to write their own alibi. As Joe Friday used to say on the start of the *Dragnet* TV crime shows:

> For every crime that's committed, you've got three million suspects to choose from. People who saw it happen—but really didn't. People who don't remember—those who try to forget. Those who tell the truth—those who lie.

This is why I have become convinced that the real battle for mass action will not be won through enemy narratives and that we need to find narratives based on cooperation, mutual interests, and our common humanity.

This is not, of course, to ever suggest that those who obstruct political action or deliberately distort the science should be let off lightly or left unchallenged. Oil companies are not just passive energy providers, whatever they like to say. They actively interfere in the political process to protect their interests. However, neither are we blameless dupes. We willingly avail ourselves of their products and the extraordinary lifestyles they enable.

This poses a challenge for generating political change. Change requires social movements. Social movements require physical targets or a product that can be boycotted, blockaded, or occupied. And a narrative of opposition requires an opponent. As Bill McKibben argues, "Movements require enemies," and in his view, this is the fossil fuel industry, which he describes as "Public Enemy Number One to the survival of our planetary civilization."

Rabbi Arthur Waskow converts the same conflict into a biblical context. Citing the biblical plagues that fell on the Egyptians, he says that "today the Pharaohs are giant corporations: big coal, big oil, and big natural gas." He adds, "The only way to deal with a modern-day Pharaoh is to organize the people."

But these targets are not *the enemy* and the struggle against them is not the place where climate change will be decided. They are *an obstacle*, and this is an important distinction. It is a delicate balance—one that mature campaigners like McKibben fully recognize.

Other struggles also need to be recognized. Gill Ereaut, founder of the communications consultancy Linguistic Landscapes, argues that narratives do not have to have an enemy—in fact many mythological tales are constructed around a quest, a challenge, even overcoming an argument, an idea, a weakness, or a way of thought.

She draws on the psychoanalytic theory of Carl Jung to suggest that if there is an enemy, it is really our "shadow"—our greedy internal child whom we don't wish to acknowledge or recognize and who compels us to project our own unacceptable attributes onto others. While climate change develops some lukewarm narratives of guilt, there are none, as I argue later, that really invite us to accept our personal responsibility.

The veteran ABC journalist Bill Blakemore, who has done more than anyone to get climate change onto American television screens, is also convinced that the real story lies in our flawed psychology. There has, he tells me, been "a grave failure of professional imagination about how to advance this great and transformative story" which should never have been "shoveled into the environmental slot."

Blakemore spent most of his working life as a war correspondent, which, one might think, would dispose him to see it in terms of competing sides or national interests. But, he points out, there "are no borders in the constantly swirling air or in the oceans" and for Blakemore the real story is about our fear, denial, and struggle to accept our own responsibility. As he says, "Climate change isn't the elephant in the room; it's the elephant we're all inside of."

Every campaign defines the language and battle lines that will determine our future thinking. If our founding narratives are based around enemies, there is no reason to suppose that, as climate impacts build in intensity, new and far more vicious enemy narratives will not readily replace them, drawing on religious, generational, political, class, and nationalistic divides—especially in the Middle East, where water scarcity could catalyze bitter conflict along religious lines. History has shown us too many times that enemy narratives soften us up for the violence, scapegoating, or genocide that follows.

THIS IS A GOOD TIME to pause and sum up the book so far.

I have already found that we interpret the world in the light of our recent experience and our attitudes. Our confirmation bias then leads us to seek out further information to confirm those existing views, or to reject information that challenges them. I found that Tea Party activists, who prided themselves on their independence, had become so distrustful of outside views that all of their information sources reinforced their existing positions.

I found that we feel compelled, by our desire to fit into our social group, to take our cues for what we should think or do from the people around us. In communities recovering from weather-related disasters I found that their understanding of the event was mediated through the cues contained in shared stories. They were entirely capable of ignoring the role of climate change because it did not fit with the stories they chose to tell.

I also found that our desire to conform leads us to exaggerate the differences between ourselves and people in other social groups. If an attitude toward climate change becomes strongly associated with a group that we actively distrust, then the science can become "polluted" by this conflict. I suggested that campaigners for and against action on climate change were equally bound by this mechanism, and even used the same language and metaphors when describing the other side.

It is already clear that the way we relate to climate change cannot be readily condensed into a simple formula of cause and effect. Our views are constantly being shaped through the negotiation between our own identity, our group loyalty, and our relationship with wider society. We are active participants, at every stage, influencing those around us as much as we are influenced by them.

The best metaphor I can find for this lies in the way that climate scientists chart the flows within the global energy and carbon systems. In their models each part of the flow is interlinked with the other parts, such that a change in one part may spread and then amplify its impacts through what scientists call positive feedbacks. There are many such social feedbacks operating in our attitudes to climate change—such as the bystander effect or false consensus effect—that exaggerate small differences and widen the divides between people.

But this can only be a partial answer to the question of why we find it so hard to act. Those who passionately accept or passionately deny climate change have one key thing in common: They all regard it as a major threat that they need to mobilize around. But, in between these two conflicting groups, the vast majority of people find it hard to accept the importance of this issue at all. When asked, they will happily tell pollsters that they are concerned about this issue, but, as I will show, they give it little other consideration and rarely if ever talk about it.

So this returns us to the original question: Is there something *innate* in this issue that enables people to disregard it in this way? How else would it be possible for people to know that climate change is a threat but not *feel* that it is a threat?

10

The Two Brains

Why We Are So Poorly Evolved to Deal with Climate Change

THE IDEA THAT OUR EVOLUTIONARY psychology makes it hard for us to deal with climate change is widespread. Paleoanthropologist Ian Tattersall in the coda to his book *The Masters of the Planet: The Search for Our Human Origins* reflects that "we are notably bad at assessing risk. Inside our skulls are fish, reptile and shrew brains." This, he says, is why we can ignore climate change and think that we won't have to face its consequences. Professor Paul Ehrlich, the outspoken population biologist at Stanford University, argues that we cannot deal with climate change because "the forces of genetic and cultural selection were not creating brains capable of looking generations ahead."

Evolutionary psychology is much contested, debated, and fought over on political and ideological grounds. Climate deniers argue that it underestimates the speed of evolution and that constant environmental and climate changes during evolutionary history have actually left us remarkably well adapted and prepared for the changes of the modern world.

Daniel Gilbert, a professor of psychology at Harvard, disagrees. He tells me that climate change is "a threat that our evolved brains are uniquely unsuited to do a damned thing about." Gilbert has given this some thought: He is an expert, and now a bestselling author, on the psychology of happiness, and he has the kind of free-roving hyperactive mind that is fascinated by everything.

Gilbert argues that our long psychological evolution has prepared us to respond strongly to four key triggers that he neatly summarizes with the acronym PAIN:

Personal: Our brains are most highly attuned to identifying friends, enemies, defectors, and human agency.

Abrupt: We are most sensitive to sudden relative changes and tend to ignore slow-moving threats.

Immoral: We respond to things that we find to be indecent, impious, repulsive, or disgusting.

Now: Our ability to look into the future is one of our most stunning abilities, but, he says, it is "still in the early stages of R&D."

As Gilbert sees it, the problem with climate change is that it doesn't trigger any of these. Of the four, he is most inclined to emphasize the lack of Abrupt and Now, which "are things that even a rabbit understands." But he would not underestimate the importance of Immoral. While we recognize that climate change is bad, it does not make us feel noxious or disgraced. He adds, "If global warming were caused by eating puppies, millions of Americans would be massing in the streets."

Unless, I suggest, Americans already eat puppies. The taboo is socially constructed, and one could readily imagine an alternative culture in which, following the lead of the hungry pilgrim settlers, it was roast puppy that had become the centerpiece of the Thanksgiving table.

Gilbert concedes the point but not the distinction. "To me," he says, "socially constructed and evolutionary are different ways of spelling the same thing. The most interesting thing about us as a species from an evolutionary standpoint is not really our opposable thumb or our ability with language; it's our social life." It is this that determines the social cues, norms, enemies, and in-group, out-group dynamics that, as I have already argued, are so important in shaping our response to climate change.

Gilbert draws on a large body of research, some of it his own, in the field of evolutionary psychology. The founders of modern evolutionary psychology, Leda Cosmides and John Tooby, like to say that "our modern skulls house a Stone Age mind," which developed to address the specific threats in what they call the "environment of evolutionary adaptedness."

In this primeval environment, the main avoidable risks were in our immediate surroundings, and it was this that led us to give such a high priority to proximity and certainty in our judgment of risk. This also makes us innately conservative and defensive of our current circumstances—what cognitive psychologists call our *status quo*. After all, survival prospects are poor for an animal that is not suspicious of novelty.

Cosmides and Tooby describe the brain as being a "Swiss Army knife" containing specialized tools designed to deal with different tasks. Thus, they argue, we are relatively poor at dealing with large issues (like climate change) and can engage with them only by breaking them down into individual tool-oriented tasks.

Like a Swiss Army knife, the brain also contains things that you never need or don't even know what they are for. Evolutionary psychologists call these "exaptations": behaviors that may have been selected in the past for completely different reasons and become co-opted into their present role following a change in environmental circumstances.

Without entering into the intense debate surrounding exaptation, suffice it to say that it contains a highly relevant core concept: that we apply to climate change the psychological tools we have evolved to cope with previous challenges, and that these may turn out to be inappropriate for this new threat. The in-group loyalties and defensiveness that evolved to support small hunter-gatherer groups may be an obstacle when dealing with a universal shared threat. As I suggest later, our avoidance of the issue of climate change may be driven by still-deeper mechanisms evolved to cope with our fears of death.

But of greatest relevance to our decision making around climate change is the discovery that this long evolutionary journey has led us to develop two distinct information processing systems. One is analytical, logical, and encodes reality in abstract symbols, words, and numbers. The other is driven by emotions (especially fear and anxiety), images, intuition, and experience. Language operates in both processes, but in the analytic system, it is used to describe and define; in the emotional system, it is used to communicate meaning, especially in the form of stories.

Brain scanning has confirmed that these systems are built into the physical architecture of the brain—the former in the cortex and posterior parietal cortex, the latter in the amygdala at the base of the brain. The neuroscientist Joseph LeDoux argues in his book *The Emotional Brain* that, as our analytic systems evolved, the amygdala was allowed to maintain its dominance in decision making because of its ability to rapidly assess threats. So, while the analytic system is slow and deliberative, rationally weighing the evidence and probabilities, the emotional system is automatic, impulsive, and quick to apply mental shortcuts so that it can quickly reach conclusions.

There has been a very strong public interest in these findings in recent years and many attempts to name them. Seymour Epstein, who first identified them as two parallel systems, called them analytic processing and experiential processing. Others call them enlightenment reason and real reason, or the reflective system and the automatic system, or System 1 and System 2. I find it easier to call them the *rational brain* and the *emotional brain*. These are not ideal names but they are easy to follow.

One of reasons there have been so many different attempts to name them is that the systems are not separate and isolated but rather in constant communication. Attempting to capture this relationship, Jonathan Haidt, a psychologist at New York University, hit on the image of an elephant and a rider. The rational rider does his best to steer the emotional elephant. He appears to be in control, though, in reality, a six-ton elephant is going to have the last say.

It is a nice image, and the second of many metaphorical elephants to appear in this book. However, this, too, is not entirely satisfactory because it underplays the communication between the two. The research shows that our rational rider will try to convince our emotional elephant and will deliberately shape arguments into stories and images that will appeal to the elephant. And the elephant is no dullard either. It is extremely adept at creating elaborate intellectual rationalizations for the rider to let it go on a path it had already decided to take. The image suggests the rider sitting under a little tent, pulling the reins, but the reality is more like Tarzan riding bareback and talking elephant (perhaps explaining why Hollywood thinks that the African jungle is full of Asian bananas).

Our perception of risk is dominated by our emotional brain. It favors proximity, draws on personal experience, and deals with images and stories that speak to existing values. As I will show later, threats that conjure up strong images or that are communicated in personal stories have disproportional sway over our decision making.

However, because the emotional brain is poorly suited to dealing with uncertain long-term threats of the kind that constitute climate change, the rational brain sometimes actively intervenes, using its abstract tools of planning and forward thinking. Indeed, experiments show that people deliberately enable this process by making an issue more distant in order to see it in rational perspective and then developing the short-term goals that give it emotional proximity. It is like a little dance—moving far away to admire your partner and then moving in close enough to kiss.

And this is exactly what we do with climate change, both personally and culturally. The theories, graphs, projects, and data speak almost entirely to the rational brain. That helps us to evaluate the evidence and, for most people, to recognize that there is a major problem. But it does not spur us to action. The divide between the rational brain and the emotional brain is embedded in the historical boundaries between science, the arts, and religion, and it is a particular risk for an issue that originates strongly in just one cultural domain—as climate change does with science—that finds it hard to engage our entire cognition. The view held by every specialist I spoke to is that we have still not found a way to effectively engage our emotional brains in climate change. Even if the rider is fascinated by the article in *Scientific American*, the elephant has wandered off looking for a banana.

So, advocates for action on climate change have to do everything they can to speak to both. They need to maintain enough of the data and evidence to satisfy the rational brain that they are a credible source. They need to translate that data into a form that will engage and motivate the emotional brain using the tools of immediacy, proximity, social meaning, stories, and metaphors that draw on experience. Every piece of climate change communication from the National Academy of Sciences to a direct-action protest outside a power station is an experiment in the alchemy of turning base data into emotional gold.

Those opposing action are playing the same game but working backward. They begin with the arguments that can appeal to the emotional brain based around the values, concerns, and emotional triggers of their audiences. They then seek the data and evidence to support these arguments, because, like the advocates, they need to satisfy both the emotional *and* the rational brains of the people they want to convince. Of course, they don't see it like this. They are convinced that they have built their emotional argument on the back of a rational evaluation of the data. And so it seems to them.

The people in between are not passive in this process either. They, too, are deliberately making a calculation about how they wish to interpret these arguments, knowing very well that if their emotional brain becomes too involved, they are likely to feel anxious and worried. As I argue later, they tend to adopt a position of wait and see: Their rational brain is sufficiently aware that they know there is a problem; their emotional brain is sufficiently engaged that it is looking out for social cues about how they should respond. And both of their brains are sufficiently detached that they do not have to deal with the problem unless actively compelled to do so.

11

Familiar Yet Unimaginable

Why Climate Change Does Not Feel Dangerous

FIVE YEARS AGO, WHEN I was living in Oxford, England, a cell phone company applied for planning permission to install a cell phone tower on the side of the local pub. The area was full of liberal professionals of the kind that congregate in university cities. When prompted, they would agree unanimously that climate change was a serious problem that someone really should do something about . . . sometime. Otherwise they really didn't think about it.

And yet the threat of the cell phone tower galvanized them into immediate personal action. Within a week of the application, two hundred people gathered in the local school hall to express their resistance to the tower, which, they said, was going to spread microwave radiation across the school playground. Some were determined to lay their bodies down in front of the installation van, if necessary.

There are some interesting similarities between the issues of climate change and cell phone towers. Both threaten uncertain impacts that are drawn out long into the future. And in both cases we contribute, through our consumption choices, to the problem we decry. My neighbors in the audience that night all had cell phones in their pockets. In retrospect, I regret not having called them up during the meeting, just to see what would happen.

However, there is one important difference: Climate change poses a vast and unparalleled threat. Cell phone towers, though, are virtually harmless. Applying even the most cautious estimates, it would take more than seventy thousand of these towers to generate enough microwave radiation to cause any health problems. This experience found me asking why highly educated people would become so agitated about an intangible and unproven risk like cell phone radiation and yet be oblivious to the equally intangible yet far-better-proven risk of climate change.

Paul Slovic, professor of psychology at the University of Oregon, is well positioned to answer this question. Slovic is the world's leading expert on the social amplification of risk. He is also a modest, soft-spoken man who might be shy of such accolades. It is, though, impossible to find a single research document on the topic that is not peppered with references to his work.

Slovic faced an uphill struggle to persuade scientists that our perception of risk is socially formed and to overcome their prejudice that, in his words, social science was soft and squishy. It was the issue of radiation—and in particular the question about why people were so much more concerned about nuclear power than they were about the dangers of medical X-rays—that launched his career in the 1970s.

Slovic identified two main drivers of risk perception: a sense of powerlessness in the face of involuntary and catastrophic impacts, which he called dread risk, and an anxiety that comes from the uncertainty of new and unforeseeable dangers, which he called unknown risk. Dread risk is reinforced by being intergenerational and irreversible. Unknown risk is emphasized by being invisible and unprecedented. Radiation is so feared because it involves both types.

Through social testing, Slovic mapped a wide range of threats against these criteria for dread and unknown risk. Chemicals, food additives, and microwave ovens score highly for their unknown risk. Nuclear weapons and nerve gas accidents score highly for their dread risk. The more mundane dangers of bicycle accidents, indoor smoke, alcohol, and home swimming pools have low scores by both criteria even though they are all major sources of fatalities.

Slovic's research explains all too well why my friends and neighbors could become so agitated about a cell phone tower. It contained a near-perfect mixture of threats: a new technology, dread risk fears of radiation, a threat to our children as they played innocently in their school playground. It also had exceptional proximity: visible, local, immediate, and with a clearly defined deadline. Finally, the coup de grâce: It had an external enemy, the faceless T-Mobile phone corporation, which had, for its own nefarious reasons, disguised this dangerous radiation-emitting tower as a flagpole.

So I ask Slovic where climate change would sit on his scales and why it is not capable of raising the same level of concern. After all, it is also catastrophic, irreversible, new, related to technology, threatening to children, and it makes people feel powerless. Surely, I suggest, this is a royal flush of both dread and unknown risks.

Slovic is not persuaded. He fully accepts that climate change is a massive problem. Indeed, he says he would work on it himself but he is now specializing in genocide and "only works on one impossible problem from hell at a time."

However, he says, it does not *feel* threatening, and that is the critical distinction. People's resistance to nuclear power, toxic chemicals, or vaccination tends to emerge at the point when something is about to change: when they take their child for a vaccination or when a nuclear plant (or cell phone tower) may be placed in their neighborhood.

But once things are accepted into our status quo and assumed to be part of normal life, it requires a far higher level of threat to have them removed. People might very well mobilize against a new energy technology that causes climate change, but not against the cars, planes, and power plants that are already woven into the fabric of their lives.

Slovic argues that extreme weather events, even highly visible ones such as Hurricanes Katrina and Sandy, are also part of our accepted way of life—our status quo—in ways that can lead us to accept rather than resist them. He suggests that extreme weather events seem familiar, and we are accustomed—in the developed world at any rate—to regard them as manageable. "Even when they do happen to us," he says, "the storm goes over. You look out the window and, hey, it's a beautiful day." As I found in Bastrop and New Jersey, people are initially traumatized but dust themselves off and focus on reconstruction and moving forward.

In language theory, the term "false friends" recognizes the trap posed by words that look and sound the same but have developed entirely different meanings—as anyone shopping for clothes on the other side of the Atlantic will find when they ask for pants, knickers, vests, or jumpers. Climate change has plenty of linguistic false friends and, as I will show, there is endless potential for misunderstanding scientific terms when they are used in a wider context. But the weather is also a kind of false friend: It looks and feels familiar, and we have a wide range of available experience to draw on that can mislead us.

Paul Slovic suggests that the third major problem is that climate change is not readily imaginable. "With threats of graphic imaginability, such as terrorism after the 9/11 attacks, you lose all sense of proportion and respond with high alarm to low probabilities. The availability bias that draws on recent experience keeps the threat alive, and the uncertainty of when the next attack might come does not diminish that fear; it amplifies it."

But because climate change does not have the same stigma, and extreme weather events have a degree of familiarity, the uncertainty of its impacts do not instill dread but rather, Slovic says, "give you the leeway to believe what you want to believe."

Believe what you want to believe? This is a telling phrase. Slovic is saying that even though it involves so many of the characteristics of dread and unknown risk, climate change does not *feel* frightening unless you actively *choose* to see it that way. If you are already inclined (by your values, politics, or social group) to see climate change as dangerous, then it looks really dangerous. If you are not inclined that way, then it looks exaggerated. Once again, the perception of climate change is being determined by the social lens you see it through, and, once again, there is a powerful feedback that tends to pull people apart.

12

Uncertain Long-Term Costs

Why Our Cognitive Biases Line Up Against Climate Change

"THIS IS NOT WHAT YOU might want to hear," says Professor Daniel Kahneman. "I am very sorry, but I am deeply pessimistic. I really see no path to success on climate change."

I assure him that this indeed what I want to hear and the reason why I wanted to talk to him. Kahneman, after all, received a Nobel Prize for his pioneering work on the psychology of decision making, and his bestselling book, *Thinking Fast and Thinking Slow*, has been a major influence on my own thinking.

In our packed cafe in downtown New York, the background noise is painfully high, and Kahneman delivers his argument a few words at a time in between long pauses for another spoonful from his seemingly bottomless bowl of tomato soup. Piece by piece, he meticulously outlines the reasons why he thinks that climate change is a hopeless problem and why it doesn't have the necessary characteristics for seriously mobilizing people's sense of threat.

His concerns are threefold. First, climate change lacks salience—by which he means the qualities that mark it as prominent or demanding attention. Like Daniel Gilbert, Kahneman argues that the greatest salience belongs to threats that are concrete, immediate, and indisputable—for instance, a car out of control driving right at you. By contrast, climate change is, he says, abstract, distant, invisible, and disputed.

The second problem, he notes, is that dealing with climate change requires that people accept certain short-term costs and reductions in their living standards in order to mitigate against higher but uncertain losses that are far in the future. This is a combination that, he fears, is exceptionally hard for us to accept.

Third, information about climate change seems uncertain and contested. As long as that remains the case, he says, "people will score it as a draw, even if there is a National Academy on one side and some cranks on the other."

"The bottom line," Kahneman says, "is that I'm extremely skeptical that we can cope with climate change. To mobilize people, this has to become an emotional issue. It has to have the immediacy *and* salience. A distant, abstract, and disputed threat just doesn't have the necessary characteristics for seriously mobilizing public opinion."

This combination of short-term and long-term decision making under conditions of uncertainty is the essence of Kahneman's lifework. During his twenty-year collaboration with the psychologist Amos Tversky, Kahneman challenged the prevailing economic assumption, called utility theory, that choices are made with a rational evaluation of future benefits.

They argued instead that our decisions are more likely to be directed by a set of inbuilt and largely intuitive mental shortcuts—what they called cognitive biases. Biases help us to apply our previous experience to new information, enabling us to decide what to heed and what to ignore. They are an invaluable tool when dealing with simple day-to-day decisions but can generate serious systematic errors when applied to complex decision making. Kahneman and Tversky found that people are consistently far more averse to losses than gains, are far more sensitive to short-term costs than long-term costs, and privilege certainty over uncertainty.

Kahneman sees climate change as a near perfect lineup of these biases. An issue is challenging enough if it concerns only losses and no gains. And it is challenging if those losses are long-term not short-term. And it is challenging if it has substantial uncertainty. Climate change appears to be the perfect combination of all three factors. I will examine each of them, in turn, in the following chapters.

I ask him whether these cognitive barriers could be overcome if people understood them better—this is, after all, one of my hopes for this book. Professor Kahneman pauses for another contemplative spoonful of tomato soup. "Actually," he says, "I'm not very optimistic about that either. No amount of psychological awareness will overcome people's reluctance to lower their standard of living. So that's my bottom line: There is not much hope. I'm thoroughly pessimistic. I'm sorry."

13

Them, There, and Then

How We Push Climate Change Far Away

POLITICIANS CONSTANTLY DESCRIBE CLIMATE CHANGE as a long-term issue and a threat to future generations. In his keynote speech on climate policy in June 2013, President Obama spoke of how we have to be caretakers of the *future*, stand up for the *future*, look after the *future*, not fear the *future* but embrace the sustainable energy *future*. Christine Lagarde, head of the International Monetary Fund, is more blunt—coming from the land of French cuisine, she fears that "future generations will be roasted, toasted, fried, and grilled."

The public duly regards climate change in the same light. In surveys, the most revealing answer comes when people are asked whether they think that climate change will affect them or future generations. In both America and Britain they give the same answer: A large majority (usually around two-thirds) say that it will *not* affect them personally. And a large majority—often of exactly the same size—say that it *will* affect future generations.

Time is just one aspect of salience. It is an innate feature of our mental categorizing that we define things in terms of their closeness: prioritizing the things that affect us, here and now, and disregarding those that affect others, there and then. In experiments, people tend to amplify this bias by deliberately choosing to regard something that is distant in one aspect as distant in other ways too.

People's perception of the risk posed by climate change duly ratchets up in steady increments the further away its victims lie. With each remove, it becomes more hazardous: first for other members of their family, then for their community, then for other Americans, then for other people in rich countries, then for poor people abroad, then for other species, and finally—the most distant category of all—for people in the future.

This tendency to distance potential impacts works in close partnership with another bias identified by Daniel Kahneman and Amos Tversky: the tendency for people to assume that they face lower risks than others do. This "optimism bias," as they named it, has been found in many different situations: people believing that other smokers are more at risk of a heart attack than they are, that other housing estates have more crime, that other drivers are more likely to have an accident, and, as I have already mentioned, that the next big hurricane will hit somewhere else.

This also applies to the environment. There is a near-universal belief that the environment is in better condition in one's own area—indeed, in a study in eighteen countries, people in sixteen of them were convinced that they had the best environmental conditions.

Climate change has other timing problems. Daniel Kahneman argues that when impacts come in intervals—such as business cycles—people's availability bias leads them to focus on the most recent event and miss the longer trend. Like Slovic, Kahneman is concerned that each successive extreme weather event then becomes accepted into our status quo and become the new baseline against which we measure change. A heat wave or flood is judged against the level set in the last heat wave or flood and we may not notice the overall scale of change over the longer term.

This is why lesser problems that deliver a single exceptional impact at a predictable moment can galvanize a far higher level of attention. Take, for example, a threat that combined an unprecedented technological cause (what Paul Slovic would call an unknown risk) with a precise and deeply symbolic timing: the prediction that the world's computer systems would collapse when the date changed on New Year's Eve 1999.

I remember vividly my local bookshop building an entire display of the opportunist books warning of the coming social collapse from the "Y2K computer time bomb," after which law and order would break down, starving mobs would roam the streets, and, as the antinuclear campaigner Helen Caldicott warned, accidental missile launchings could lead to "Armageddon."

And, of course, nothing happened in either the United States, which had poured an estimated $134 billion into the pockets of software analysts and computer programmers, or in South Korea, Italy, or Ukraine, countries that had done next to nothing. On January 1, a few slot machines and cash registers became momentarily confused, that was all.

Campaigners have always struggled with climate change's un-engaging timeline and tried to find ways to generate the same compelling sense of urgency and symbolism as Y2K. In 1947, the *Bulletin of Atomic Scientists* had hit on the novel image of the Doomsday Clock, which is always close to striking midnight, to dramatize the risks of nuclear weapons. In 2012, the hands were moved to five minutes to midnight to recognize the coming climate change catastrophe, and the following year, Rajendra Pachauri, the head of the Intergovernmental Panel on Climate Change, announced its latest report with the words "We have five minutes before midnight."

Environmental campaigns around climate change also manufacture deadlines to create urgency. Back in 1990, I was an intern in the smoke-filled offices of the *Ecologist* magazine as the editorial staff pulled together a new book warning of environmental collapse with the title *5000 Days to Save the Planet*. The deadline ran out in the same year that the British think tank Institute for Public Policy Research released its own deadline report: "Ten Years to Save the Planet."

And the baton kept being passed on. In 2007, an international team of celebrities united under the banner Global Cool said that ten years was the deadline to "save the planet." That year, the World Wildlife Fund warned that five years was the "small window of time in which we can plant the seeds of change." Each successive climate conference provides a countdown, and I am already receiving action alerts on the run-up to the Paris conference in 2015 telling me how many months we have left to save the world.

In 2008, the London-based New Economics Foundation launched the campaign 100 Months to Save the World. Prince Charles became a strong supporter of the deadline, declaring shortly after, from a podium in his modestly appointed pad in St. James's Palace, that there were now only ninety-six months left to reverse what he calls "de-souled consumerism" and that the "age of convenience" was over.

But it was not over at all. Charles may have converted his fleet of cars to run on wine-based ethanol, but in every other respect, life appears to be largely unchanged. I ask Andrew Simms, who invented that campaign slogan, what happens in 2016 when, as Slovic says, we open up the windows and find that it's still a beautiful day. "Of course," Simms says, "we never said that the sky would set fire and all forests burn. Really what happens is that it becomes more, rather than less, likely that we go over the safe threshold and positive feedbacks kick in."

And here lies the problem. Without an explanation, the deadline can seem to be arbitrary and imposed. However, as soon as it has an explanation, the campaign is immersed back into the tangle of probabilities, uncertainties, and cost-benefit analysis that it originally sought to avoid. No one is ever going to march under a banner of "100 Months Before the Odds Shift into a Greater Likelihood of Feedbacks," especially given that many experts fear that we may have already crossed that line.

But even if climate change has no pressing deadlines, this does not mean that the *anticipation* of future threat will not motivate us and enable us to overcome any tendency we have to discount the future.

For the past twenty years, George Loewenstein, professor of psychology at Carnegie Mellon University, has been exploring the effect of anticipation on our attitudes to future loss. Fearful anticipation is, he says, highly motivating. He agrees that we do tend to discount the future but feels that it is "easy to exaggerate its importance." His research suggests that the anticipation effect is likely to be even stronger in a case, like climate change, in which people anticipate a deteriorating condition, what he calls a "decline sequence."

Anticipation was, for example, a major factor in people's fear of nuclear war, another issue whose future impacts face great uncertainty. Risk specialists talk of the signal value of individual incidents, such as nuclear accidents, which portend future disaster. In this sense extreme weather events offer a string of signal events that feed the anticipation of future loss, which explains why climate disasters can heighten concern among those already disposed to accept the threats of climate change.

However, Loewenstein stressed to me, people's primary response will always be to mitigate the dread, even if this means avoiding the problem itself. There is, he warned, a very narrow boundary between not believing that the problem is happening at all and being so afraid that you are immobilized. So, he says, "from a psychological point of view, the crux— the key issue—is our capacity to generate our own arguments that can interfere with that fear response." These arguments include our capacity to, in Slovic's phrase, "believe what you want to believe."

So there is a legitimate question to ask: Given our ability to deliberately construct a narrative that can reduce our sense of fear, is climate change really a distant issue for future generations, or have we just decided that we want to see it as one?

Certainly there is nothing new about climate science. The science of the greenhouse effect goes back to the theories of Joseph Fourier in 1824. In 1896 the Swedish chemist Svante Arrhenius first calculated the impacts of doubling the amount of carbon dioxide on global temperatures. He estimated that it would raise global temperatures by five to six degrees Celsius, well within the range that climate scientists, with their powerful computer models, now fear is in the offing.

Nor is there anything new about climate change as a political concern. The first major political warning emerged in 1965 when President Lyndon B. Johnson's Scientific Advisory Council cautioned that the constant increase in atmospheric carbon dioxide could "modify the heat balance of the atmosphere." In 1992, every nation in the world signed the U.N. Framework Convention on Climate Change, and an entire generation has grown up in its wake.

The impacts of climate change are hardly in the future either. I write this after the 350th month in a row that is warmer than that month's historical average. Although temperatures vary from year to year, the trend is also very clear.

Scientists, though, always stress the importance of natural variability in climate systems and only start to express confidence in their models in the time horizon that most people see as being beyond their immediate concerns—typically 2050, a date that researchers have found to be set so far in the future as to be "almost hypothetical" for the general public.

Politicians are all too happy to talk about climate change in these terms so that they can postpone difficult decisions as far into the future as possible. Western governments, in particular, wish to look forward and have no desire to remind anyone of their historical responsibility for two hundred years of emissions from industrial growth and land clearance.

So climate change is a future problem. But it is also a past problem and a present problem. It is better thought of as a developing process of long-term deterioration, called, by some psychologists, a "creeping problem." The lack of a definite beginning, end, or deadline requires that we create our own timeline. Not surprisingly, we do so in ways that remove the compulsion to act. We allow just enough history to make it seem familiar but not enough to create a responsibility for our past emissions. We make it just current enough to accept that we need to do something about it but put it just too far in the future to require immediate action.

14

Costing the Earth

Why We Want to Gain the Whole World Yet Lose Our Lives

ARGUMENTS AROUND CLIMATE CHANGE HAVE always been based on the short-term and long-term costs. Advocates of action tend to emphasize the long-term costs of inaction and, taking a leaf from the book of health campaigners, stress that there are immediate economic benefits from moving to a sustainable economy. Opponents of action play the same game in reverse, emphasizing the short-term costs and the painful disruption to the status quo while playing down the long-term costs of climate change impacts, or rejecting the existence of the problem altogether.

When the issue first emerged in the early 1990s, arguments for government action were often phrased in terms of a cost-benefit analysis. In retrospect, these read as overly rationalist and almost deliberately evasive of any emotional or ethical component. One influential report from 1993 argued that "a comparison between the costs of greenhouse prevention and the benefits of avoided warming is the backbone of an economically rational greenhouse response." Another, released around the same time by the British Royal Society, argued for an "optimal level of safety where the extra cost of any extra reduction [in emissions] just equals its benefits but goes no further."

In 2006 Sir Nicholas Stern, the former chief economist of the World

Bank, wrote the hugely influential *Review of the Economics of Climate Change*, which claims—though his critics argue otherwise—to provide a balanced economic evaluation of the short- and long-term costs of climate change. From his seven hundred pages of dense analysis, Stern produced a key finding: that reducing the greenhouse gas emissions that cause climate change will cost us around 1 percent of our annual income for the next fifty years. However, he warned, if we do not do this, the annual costs from extreme climate events will rise to anything from 5 percent to 20 percent of our annual income—indefinitely.

This sounds like an unassailable argument for immediate action—and indeed it would be if we evaluated costs in the same way that we weigh up benefits. However Daniel Kahneman's research found that we are far more prepared to take risks on losses than gains, especially if the losses lie in the future.

For example, to paraphrase one of Kahneman's experiments, imagine that you have been offered nine hundred dollars. Would you be prepared to take a 90 percent chance of boosting it to one thousand dollars with a 10 percent chance of losing it? The majority of people say no. Imagine, then, that this is a loss and you are required to pay nine hundred dollars. Would you be prepared to accept a gamble that offered a 10 percent chance of being let off the full amount, but with the 90 percent chance that your debt would rise to one thousand dollars? This seems more tempting even though the actual condition is exactly the same. The large majority of people would take this chance.

Research has also found that people give an overwhelming priority to the short term over the long term and that they discount the future—*discount* here being an economic term for reducing value over time. When results for different time scales are plotted on a chart, the resulting curve is hyperbolic—that is to say that the sense of relative loss is most acute in the near future but declines in the distant future. Hence—not too surprisingly—this is called hyperbolic discounting.

Taken together, this research relates directly to climate change, suggesting, as Kahneman feared, that people will be strongly disposed to avoid short-term falls in their living standard and to take their chances on the uncertain but potentially far higher costs that might come in the longer term.

As predicted by the hyperbolic discounting model, governments have proven to be extremely unwilling to incur costs in the short term but perfectly willing to accept far greater costs in the future. The governments of the European Union, the U.S. state of California, and the Canadian province of British Columbia have all declared a long-term target of reducing emissions by 80 percent within forty years. So far, they have managed to achieve a meager half a percent reduction per year.

The larger measures that might achieve the target are safely placed further down that precipitous slope of parabolic indifference, always just over the horizon: sometime soon, but not quite yet, and, as skeptics like to point out, surrounded by a remarkable vagueness about what they might cost. Much recent climate change policy has been based on treading water—what the U.S. National Academy of Scientists calls a "muddling through" strategy.

Clearly, rational cost-benefit analyses do not stimulate a sense of threat or motivate action among policy makers. If anything, they seem to encourage them to take a chance. When Sir Nicholas Stern offers them a choice between a certain one percent decrease of income now or a 5 to 20 percent decrease of income at some point in the future, it feels disconcertingly like one of Daniel Kahneman's experiments in temporal discounting.

What is more, senior politicians and business leaders are inveterate risk takers who have been successful through a string of lucky gambles, which they prefer to credit to their unique talents. Inviting such people to take a gamble on the future is like inviting an alcoholic to pour himself a drink.

But the more important question in trying to understand why people are so unwilling to accept climate change is whether humans as a whole are *innately* disposed to disregard any threat that requires sustained payments in order to avoid greater, but less certain, long-term losses.

Clearly, this is not necessarily the case. Each year people and businesses pay more than $4.3 trillion in insurance premiums as a protection against uncertain future risks. In the year of Hurricane Sandy, insurance payouts for climate-related damage soared to a record $44 billion, with an additional $21 billion from the U.S. National Flood Insurance Program.

Even greater costs are involved for military defense. Every year governments around the world justify $1.7 trillion in military expenditure in order, they claim, to protect citizens against entirely uncertain and ill-defined threats. Since 2001 American politicians have approved $300 billion in federal government spending to defend the United States against further terrorist attacks. This profligacy in response to an entirely uncertain and unquantifiable risk has the support of the large majority of

Americans, including many people who, in other regards, would strongly oppose government spending.

Again we are reminded that the cognitive biases that appear under the tightly controlled and artificial conditions of a psychology experiment are, in real life, subservient to people's culture, social norms, and in-group identity.

There is even a persuasive argument that the biases that reveal themselves in cognitive psychology experiments may also be culturally determined. The participants in the vast majority of experiments that built the models of cognitive biases are university students. Common sense would tell us to be wary of drawing conclusions about innate human attitudes to long term risk from overconfident young people with a marked tendency to smoke and drink too much.

A team of anthropologists at the University of British Columbia termed these experimental subjects "WEIRD," a provocative acronym for people from countries that are Western, educated, industrialized, rich, and democratic. They described them as an extremely unrepresentative and "exotic" group worthy of anthropological study in their own right. To prove their point, the anthropologists showed that archetypal psychological experiments could generate very different results when repeated in different cultures and asked for "caution in addressing questions of human nature from this slice of humanity."

And even within WEIRD countries, a cultural disposition to long-term planning can readily overrule the cognitive bias toward temporal discounting. A team based at Copenhagen Business School found little evidence of any discounting among the cooperative Danes and, as their report charmingly put it, virtually no sign of "hyperbolicky" behavior. So in Denmark, a national culture that historically has seen a high degree of cooperation around common goals can overrule the disposition to short-term interests.

The importance of cultural context applies to all other forms of cost-benefit analysis. For example, mainstream economic theory assumes that people will avoid incurring short-terms costs to themselves even though their personal action might prevent larger longer-term costs for others.

In a famous and much quoted anecdote, the Nobel laureate economist Thomas Schelling told of being caught in an hour-long traffic jam of vacationers returning from Cape Cod. When people finally saw the reason for the delay—a mattress blocking one of the lanes—no one took the initiative to stop and move it. "For all I know," he wrote, "it may still have been there the following Sunday."

Schelling argued that nobody moved the mattress because there was no system to reward them for doing so. With his tongue firmly in cheek, he suggested a market-based system "that a traffic helicopter proposes that each of the next hundred motorists flip a dime out the right-hand window to the person who removed the mattress."

Schelling then applied the same thinking to energy conservation. Even though we are urged to turn down our air-conditioning in the summer to avoid brownouts, he said, we do not do so because we know that our reduction would account for only an infinitesimal part of the total demand and therefore be of no benefit to ourselves. Without a penalty or a reward system, he argued, there is no motivation to commit anonymous acts of altruism.

Except that this is not how people in Japan behaved in 2011 after the Fukushima Daiichi nuclear power plant disaster in 2011 led to a sudden fall in the electricity baseload. When called upon to make a personal sacrifice to avoid brownouts, people willingly sweltered in indoor temperatures of over eighty degrees Fahrenheit. Without duress or reward, people willingly turned down the air-conditioning in their homes even when no one could see them doing so. The result was a 20 percent fall in peak energy demand in Tokyo throughout the summer.

So people's tendency to avoid costs and act only in their self interest—often considered major obstacles to action on climate change—can be overruled by a sufficiently strong appeal to group identity and a visible social norm. The energy crisis was given salience with posters and reminders on all media, and even large billboards above major crossings that flashed daily rates of power consumption and the likelihood of a blackout. Following the tsunami that had killed sixteen thousand people, this was a time of exceptional unity, similar to that in the United States after 9/11. People were actively seeking ways to make a personal contribution, just as they do in wartime when they are brought together in the face of a common enemy.

And there was another very relevant factor at work: *informed choice.* In Japan, the social prominence of the energy shortage led people to see the normally neutral act of turning on an air conditioner as a morally charged *choice* between self-interest and the collective good. There is a very large difference between how people respond to a risk that is an accepted part of their current condition—their status quo—and to a risk that is presented as an informed choice. An experiment by the University of Chicago economist Richard Thaler, who has frequently collaborated with Daniel Kahneman, showed this very well. Imagine that you have been exposed to a fatal illness that has a one-in-a-thousand chance of killing you. How much would you pay for a vaccination? On average, participants in the experiment were willing to pay two hundred dollars for a vaccination to remove this small but deadly risk.

Now imagine that you have been invited to be a volunteer in medical research into this same disease. You will need to be voluntarily infected, with the same chance of death—one in a thousand. How much payment would you require as compensation for this risk? If you are like me, you would refuse to take part at all. The average of those who would accept payment was ten thousand dollars.

The experiment is frequently cited as evidence that people are more inclined to accept risk if it is an unavoidable part of their status quo. But, to my mind, it says something far more interesting about the role of informed choice and the underrated importance of anticipation. When you imagine being a volunteer, you can anticipate the horror of discovering that you have, quite needlessly, *chosen* a premature death.

This experiment contains important lessons for those building action against climate change. If climate change is regarded as an unavoidable condition, like a disease that we have already been exposed to, we will become resigned to it and, at most, might pay something to reduce our exposure to future impacts, just as we pay insurance on our homes.

If, however, climate change is regarded as an active and informed choice, it feels far more like being a volunteer in the medical research. Imagine, for example, that you are offered an immediate boost in your current standard of living if you agree to pass on an irreversible disruption of the world's weather systems to your children (if you feel you can face it, look at the four-degree impacts at the end of this book to remind yourself of what is on offer). How much more income would you like to receive for that?

As soon as it is presented this way, all sorts of other considerations come into play—of anticipation, fear, responsibility, guilt, and shame. There is no option of being an innocent bystander to a crime that you have knowingly agreed to.

Climate change is never presented as a choice in this way. Most energy and fuel use is entirely automatic or woven into our daily lives. Government policy, in which decisions are more carefully constructed, deliberately removes or sidelines climate change in its choices. Even the people who deny climate change have never chosen short-term personal consumption over long-term collective climate disaster: They have chosen to believe that there is no problem.

There are, I believe, rich opportunities here for rethinking climate change in ways that might overcome the stultifying cognitive indifference to future loss: to talk less about the costs of avoiding climate change and more about the lousy deal we are getting in return for a marginally higher living standard. What is required is a moment of *informed choice* when people have to decide whether they want to accept this risk and, with it, the responsibility for being wrong.

Above all, as I will argue repeatedly in this book, people will willingly shoulder a burden—even one that requires short-term sacrifice against uncertain long-term threats—provided they share a common purpose and are rewarded with a greater sense of social belonging.

15

Certain About the Uncertainty

How We Use Uncertainty as a Justification for Inaction

UNCERTAINTY IS (I CAN SAY with some certainty) likely to be a major reason why people ignore climate change. In experiments, uncertainty about future outcomes is one of the key factors that lead people to act in their own short-term self-interest.

Policy makers and campaigners on all sides understand very well the importance of uncertainty in regard to action. This is why the U.N. Framework Convention on Climate Change expressly states, in its third principle, that a "lack of full scientific certainty should not be used as a reason for postponing measures" to minimize the causes of climate change. And this is why President George W. Bush excused his inaction on the issue by saying that "no one can say with any certainty what constitutes a dangerous level of warming, and, therefore, what level must be avoided."

The main source of public uncertainty, though, relates to the wide-spread perception that scientists are themselves divided on the issue. In the United States, more than a third of people believe that "most scientists are unsure whether global warming is occurring or not."

In part this distorted perception is due to the way that the media presents the issue as contested and adheres to a debate format that pits a climate scientist against a contrarian drawn from a spinning Rolodex of professional deniers. But this uncertainty also originates in the professional caution with which climate scientists present their findings.

There is even widespread uncertainty over the very meaning of the word *uncertainty*. The precise language of science uses the word to mean the extent to which the weight of available evidence supports a conclusion. Scientists argue that full certainty is unattainable, indeed damaging, and that the maintenance of doubt is the very *foundation* of the scientific method.

However, the lay public uses the word in a quite different way: to mean the extent to which the expert is confident in his or her stated opinion. When scientists say *uncertain*, the public hears *unsure*, and considers them less reliable or trustworthy. Thus we might be more inclined to trust an expert who is certain that something is unlikely than one who is uncertain that something is very likely.

Social trust is determined by confidence and is conveyed by body language, eye contact, and a clear and unfaltering delivery. Scientists can still generate trust in their work if they can communicate their uncertainty with social confidence. All too often, though, professional scientists project a lack of confidence, especially when set up against a professional contrarian who spends much of his life in television studios. Here, for example, is a leading climate scientist feeling the pressure in a live TV debate against a leading climate denier.

"There could be catastrophe in the air. We hear all kinds of explanations about the uncertainties, but the uncertainties—and indeed one of the quotes in the [denier's] book is 'uncertainty is uncertainty.' Well, uncertainty can play both ways."

Rather than talking confidently about the certainties of what is known, he is talking unconfidently about the uncertainties of what is not known. He is trying to say that the uncertainties are themselves dangerous, but for the listener, it seems as if he has just said "I don't know" five times in a row.

This is not to say that climate change is uncertain at all: As with any complex issue, it can be read in terms of layers of confidence. Some aspects are well known, well understood, and almost certain. Some parts are conjectural, little understood, and highly uncertain. As former U.S. Defense Secretary Donald Rumsfeld would have it, there are known knowns, known unknowns, and unknown unknowns.

Defining climate change as a whole as certain or uncertain is therefore a choice. Advocates of action focus on the known knowns and emphasize the scale of agreement around that. Opponents of action, such as the skeptical climatologist Judith Curry, emphasize the "whole host of unknown unknowns that we don't even know how to quantify."

This interpretation of climate change as exceptionally uncertain is as much a matter of confirmation bias and convenient storytelling as the sibling arguments that it is a distant future issue or that action requires unacceptable losses. Even the most skeptical of skeptics cannot, with an open mind, explore this issue and not conclude that the core science is extremely firm.

In 2008 Professor Richard Muller did exactly that and decided that climate science was in need of a more combative and *rational* challenge from a critical outsider.

Muller, I should say, likes the word *rational*. Sitting together at a bustling sidewalk cafe opposite his laboratory at the University of California, Berkeley, where he teaches theoretical physics, he tells me that the Senate was quite *rational* to reject the Kyoto Protocol. The skeptical public, he says, has been more *rational* than people give them credit for. For someone who is clearly very emotionally attached to his arguments, Muller speaks an awful lot about rationality.

Muller is extremely unwilling to accept any conventional opinion on climate change: There is, he says, not a single weather event that can be attributed to it using the standards of science; almost all solutions are pointless; the United Nations is "shameful"; electric cars are all about feeling good—they are "like the person who goes out to get exercise and then gets a milkshake."

Muller was repelled by all the talk of a climate change consensus and formed a small research team to thoroughly (and rationally) review the link between greenhouse gas levels and temperature data. Opponents of climate change expected him to explode the entire edifice of climate science. The oil-billionaire Koch brothers, whom I introduced earlier as the poster demons of climate change campaigners, supported his research generously.

Then, in July 2012, Muller and his team published their main conclusions: Global warming is real and caused by humans.

Confirming what other scientists had been saying for more than twenty years was hardly news. What was important was that, even when it had been scrutinized from a position of extreme skepticism, the scale, size, and seriousness of climate change turned out to be based on extremely strong evidence. Not *beyond* argument—for Muller, nothing is beyond argument—and there is plenty of room for energetic debate over the future impacts or the policy solutions. But the problem is only uncertain if you are determined to see it that way.

Which is, of course, what the deniers have always done. In 2002 the communications specialist Frank Luntz briefed Republican candidates: "Should the public come to believe that the scientific issues are settled, their views about global warming will change accordingly. Therefore, you need to continue to make the lack of scientific certainty a primary issue in the debate."

The public perception of the uncertainty of climate change is shaped by the undue confidence in the way that we talk about every other global threat. No one polls economists in order to obtain a level of certainty that increasing the money supply will avoid economic recession, and no analysis puts defined levels of likelihood on an evaluation of the nuclear threat posed by Iran. The application of this probabilistic language to climate change generates the specter of uncertainty, even when the scientists themselves have an abnormally high level of consensus.

Stranger still, it is exactly those politicians playing up the uncertainties of climate change who embrace uncertainty as a justification for military preparedness. Mitt Romney, the first presidential candidate to openly deny climate change, justified increasing spending for the military because "we don't know what the world is going to throw at us down the road. So we have to make decisions based upon uncertainty." Former vice president Dick Cheney, another outspoken denier of climate change, said that "even if there is only a one percent chance of terrorists getting weapons of mass destruction, we must act as if it is a certainty." Donald Rumsfeld supported this argument with another typical circumlocution: "Simply because you do not have evidence that something does exist does not mean that you have evidence that it doesn't exist. "

So a one percent chance of a terrorist attack should be acted on as though it is a certainty, but a ninety percent chance of severe climate disruption is too uncertain for action? In all of these arguments, the actual certainty or uncertainty is of little relevance and is mustered to support decisions that have been guided by political ideology.

In fact the most rational and considered response to the uncertainties of climate change can be found among military strategists. General

Gordon Sullivan, former chief of staff of the U.S. Army, regards the uncertainties of climate change as no obstacle to action at all: "We never have 100 percent certainty. You have to act with incomplete information. You have to act on your intuition sometimes."

It is now routine to include climate change as a potential threat to U.S. national security. The Center for a New American Security, whose board bursts at the seams with generals, admirals, and, typically, the acting secretary of defense, published a report in 2013 confirming that climate change is a serious national security issue. As General Chuck Wald, former deputy commander of U.S. European Command, puts it, "There's a problem there and the military is going to be a part of the solution."

Not surprisingly, this makes some observers very nervous. The liberal journalist and activist Naomi Klein has long argued that crises are exploited as a means to centralize power and subvert democracy. Her fear is "that climate change is the biggest crisis of all and it will be exploited to militarize our societies, to create fortress continents."

If this is the case, then that process has already been started by the military strategists using the language of uncertainty to justify a military response. Uncertainty is, like proximity and cost, an area within which different interest groups shape the language surrounding climate change to meet their own objectives.

People *do* doubt climate change because they perceive it to be uncertain. And this, in turn, affects their willingness to respond to it. However, as can be seen in comparison with the far greater uncertainties of other high-profile issues, language around uncertainty, like that around timing and costs, is manipulated to support the interests of those who oppose action or, in the case of the military, those aching to be in the middle of it.

16

Paddling in the Pool of Worry

How We Choose What to Ignore

EVERY DECEMBER THE LITERARY AGENT John Brockman goes through his address book and asks top scientists and writers to ponder a single question for the readers of the *New Yorker* magazine. In 2012 he asked them, What should we be worried about?

Professor Brian Knutson, a psychologist at Stanford University, replied that he was most worried about worry—or, as he suggested—he had a "metaworry about worry." I warn you that this is the first of many "meta"s you will encounter in this book.

Knutson said that our tendency to worry (and our personal disposition to do so) has been set by evolution at an optimal level because "those who worried too little died (or were eaten), while those who worried too much failed to live (or reproduce)." Unfortunately, he argued, our worry systems are entirely inadequate for coping with climate change. He did not want us to worry more—this would generate "hyperworry," which could freeze us up entirely. Rather, he said, we need to "turn our ancient worry engines in new directions."

I find that talking about *worry* can provide a more useful analysis than talking about *risk*. Risk can be evaluated and measured and engage the rational brain. But when we ask people what they worry about, we get a far stronger indication of their emotional perceptions and, as Knutson suggested, the threats they have chosen to ignore.

Patricia Linville and Gregory Fischer at Duke University argue that people's capacity for worrying about problems is limited and rationed. They have neatly named this the *finite pool of worry*. There is, they say, constant competition for space in the pool, and the modern media is always trying to get our attention by creating new emotionally charged issues to worry about. The result, Linville and Fischer argue, can be an emotional numbing—a protective indifference to issues that are not of immediate personal concern—which narrows the criteria for space in the pool, or even shrinks its total size.

So what, then, has happened to worry about climate change? This issue is a major threat, has been growing in prominence for twenty years, and has been accompanied by a string of high-profile extreme weather events. Has this enabled it to secure a corner in the pool of worry, or, as Knutson worries, do our evolutionary worry detectors refuse to grant it admission?

For the past ten years, Tony Leiserowitz, the director of the Yale Project on Climate Communication, and Edward Maibach, the director of George Mason University's Center for Climate Change Communication, have been watching the public concern about the issue rise and fall. They have been trying to identify a clear signal in all the static because, like temperatures and ice levels, public attitudes are subject to random variation, which can be confusing or hide a longer-term trend.

Certainly the short-term trend is indisputable. Across the polls, across the Western world, public concern about climate change rose steadily through the early 2000s, peaked around 2007, and thereafter went into decline, especially among people with conservative politics.

Sitting in his oak-paneled study high up in the Yale Department of Forestry and Environment, I ask Leiserowitz what he thinks is going on. He says that there were two factors coming together around 2007. The first was the deepest recession since the 1930s, and 2007, was the year in which the housing bubble popped and U.S. unemployment rates rose over 10 percent. Climate change cannot feel as salient to people as the threat of losing their job or the very visible foreclosure signs on the street.

The second issue was the virtual collapse of media coverage. In the two years following the 2009 world climate conference in Copenhagen, overall media coverage fell by two-thirds, and evening news coverage dropped by 90 percent. This was compounded by the decline in the size of the mainstream media, leading to a lower quality—and quantity—of environmental reporting.

However relevant these factors are, the truth is that concern about climate change has never been especially high at any time. For the past twenty-five years, the Gallup organization has asked people how much they "personally worry" about a variety of environmental issues. There has never been much interest in climate change; worry levels have always wobbled between "only a little" and "a fair amount," below both river and air pollution.

This is what opinion polls find when they ask people, upfront, how much they worry about climate change. About half of Americans know that when a pollster asks the question "How worried are you about global warming?," the appropriate response is to indicate some worry. However, when they are not prompted to give an answer, they scarcely mention it at all. Every year since 2001 the Pew Research Center has asked people to choose the policy issue that should be a high priority for the president. "Dealing with global warming" has never risen above the bottom slot and is probably only there at all because it was included in the list of options.

Naturally enough, as Leiserowitz predicted, the salient issues of the economy, jobs, terrorism, and health care are always on the top of the pile for the president. But far-less-salient and long-term issues such as the budget deficit, the influence of lobbyists, and even "dealing with the moral breakdown" are also regarded to be far be more pressing than climate change.

What is interesting is that none of these issues are ones over which people have personal control. It is sometimes argued that people do not accept climate change because they feel powerless to do anything about it. In the wider psychology of coping, a perceived sense of powerlessness leads to helplessness and depression. There is some research evidence that people stop paying attention to global climate change when they realize that there is no easy solution for it.

But it is clearly more complex than this. People have no personal power over terrorism or drug use or the national economy, but that does not prevent them from talking about it and demanding collective action. Ironically, through their own emissions, they may have *more* personal involvement in climate change than any of these issues, though, as we shall see in later chapters, this creates its own problems.

Nor is there any reason to suppose that if these other items were removed from the pool of worry that climate change would find room to move in, just as there is no evidence that people in countries with lower crime rates, deficits, unemployment, or river pollution than the United States have correspondingly higher levels of concern about climate change.

The pool of worry is a metaphor for cognitive processes by which we select what we wish to pay attention to, and what we choose to ignore. The past twenty years have seen a huge increase in research into the processes of attention and there is a growing consensus that such selection processes are fundamental to our thinking at every level.

According to the Canadian sociologist Erving Goffman, we manage our attention through "schemata of interpretation," which, thankfully, he also described using the far more memorable term: *frames*.

Goffman explained that frames are constructed of our values, our life experience, and the social cues of the people around us. We decide what information we wish to pay attention to—placing what is relevant, important, familiar, or rewarding to know inside the frame.

Frames are active too. They seek out, scan, and select new information. George Lakoff, a cognitive linguist at U.C. Berkeley, argues that frames have a physical presence in our brains, are embodied in our neural circuitry, and are strengthened through use. This is, he stresses, a dynamic process in which new frames build onto existing, established frames to form a coherent system.

Climate change is not a frame, but it has become framed. That is to say that people have applied their existing frames to the issue, allowing them to decide whether it is important to them and what position they should take on it. Everything we see and hear about climate change triggers frames: responsibility, resistance, freedom, science, rights, pollution, consumption, waste—all are frames with their own associations.

However, the nature of framing is that it does not just select what to pay attention to; it also selects what to ignore. Frames are like the viewfinder of a camera, and when we decide what to focus on, we are also deciding what to exclude from the image we collect. Research suggests that our ability to choose what to ignore may be just as important for our psychological functioning as our ability to choose what to attend to—and that it is this skill that enables us to cope with the information-supersaturated modern urban environment.

So far this book has focused on what is said about climate change, including the loud and intensely politicized debate and the arguments about cost, certainty, and impacts. But of equal interest is what is deliberately *not* said, and the extent to which climate change is not just marginalized but also entirely removed from the public consciousness—sitting permanently on the beach but never in the pool.

17

Don't Even Talk About It!

The Invisible Force Field of Climate Silence

I AM CONSTANTLY DROPPING THE term *climate change* into conversations with strangers. I may talk about my own work or relate it to the weird weather or some other issue that is hogging a prime spot in their pool of worry. I am very relaxed and casual about it—after all, no one wants to find herself sitting next to a zealot on a long-distance train journey.

Really, though, it doesn't seem to matter how I say it, because the result is almost always the same: The words collapse, sink, and die in midair, and the conversation suddenly changes course. It is like an invisible force field that you discover only when you barge right into it. Few people go that far, because, without ever having been told, they have somehow learned that this topic is out-of-bounds. That is why they know that if someone else inadvertently enters this zone, it is a good idea to find something new to talk about.

In America I find that the native friendliness dissipates the instant the words *climate change* enter a conversation. If I am talking to a couple, one person will continue to talk with me (it would be rude not to) while the other will instantly turn away and find some adjacent distraction.

When pressed two thirds of people admit that they rarely or never talk about it, even inside the close circle of their friends and family members. Women talk about it far less than men do, and as a group, younger women talk about it less than anyone, especially, as I will explain later, those with children. Another survey found that a quarter of people have never discussed climate change with anyone at all. In real life, it seems that the most influential climate narrative of all may be the non-narrative of collective silence.

None of this is of any surprise to Eviatar Zerubavel, a professor of sociology at Rutgers University. Zerubavel is an expert on the sociology of socially constructed silence, which, he says, is as much a part of our communication as speech, "like a substance that fills in the pauses and cracks and crannies of our discourse."

I ask Zerubavel how we can study a silence—how we can measure something that people do not recognize as being absent. He puts it this way: "We have not talked at all about zoological gardens. That is not because we are deliberately avoiding it; the subject has simply not come up in our conversation. I would call this *inattention* because we can easily explain why we have not talked about it. But *disattention* is something very different. That is when we deliberately fail to notice something and cannot even explain that silence."

"So," he continues, "what I look for is a silence about the silence—what I would call a meta-silence. The meta-silence is that we don't talk about the elephant in the room, *and* we don't talk about the fact that we don't talk about it." And that, as Zerubavel points out with glee, means that the refusal to talk about the elephant becomes an even bigger elephant. Presumably, both have escaped from the zoological gardens we have been not talking about.

My friend Mayer Hillman, a senior fellow at the Policy Studies Institute and a passionate climate change campaigner, tells a story that shows this meta-silence at work. He was attending a dinner party with retired liberal professionals like himself. People were talking about their latest holiday trips, and Mayer could not resist bringing up the issue of climate change and the impacts of their airline flights on future generations.

The room went very quiet. Then a guest decided to break the ice. "My word," she said, "what a *lovely* spinach tart." Oh yes, everyone agreed emphatically, it was a *very lovely* spinach tart. According to Mayer, they spent the next ten minutes talking about the tart, the fresh spinach, and the recipe. In Zerubavel's terms, this inability to talk about the issue or to even verbalize the reasons for not talking about it is a meta-silence.

Zerubavel cites the nineteenth-century sociologist Emile Durkheim's observation that there is a distinction between individual facts—things that you do at an individual level—and social facts, the behaviors that operate at a collective level. So, he says, "it makes a lot of sense to talk about denial and silence about climate change as operating both individually and totally differently at the collective level."

In a pioneering study, Kari Norgaard, an associate professor of sociology at the University of Oregon, set out to understand how people in a remote coastal town in Norway came to terms with climate change. In the course of her forty-six interviews, Norgaard repeatedly hit the invisible force field of silence I described. Most telling, she wrote, was that the issue often killed conversation: "People gave an initial reaction of concern, and then we hit a dead zone where there was suddenly not much to be said, 'nothing to talk about.'"

Yet people openly recognized that the weather was changing dramatically. In particular there was deep concern that the ski hill, an essential component of the town's local economy and identity, was opening later and later in the holiday season and even then only with the help of artificial snow. As I found in Bastrop and New Jersey, despite the evidence seen with their own eyes, people refused to discuss climate change. Norgaard found that it formed a detached reality—or, as a local teacher put it to her, "We live in one way and we think in another. We learn to think in parallel. It's a skill, an art of living."

Norwegians have particularly good reasons for ignoring climate change. Norway's cultural identity, Norgaard explained, is based around a mythic narrative that it is a small and humble nation that lives simply and close to nature. Norwegians pride themselves on being honest and conscientious global citizens and their government speaks often of being a world leader on climate change.

Norway is a leader all right, though not in the way it would like us to think. It is the eighth largest exporter of crude oil in the world, and its emissions grew five times faster than its already generous allowances under the Kyoto Protocol. Everyone in Norway has a direct personal stake in this oil economy, thanks to the six hundred billion dollars saved in the state oil fund, which now includes a two-billion-dollar stake in Alberta's tar sands. All in all, Norway is a spectacularly large contributor to climate change and, thanks to its egalitarian traditions, it has shared that responsibility across its entire population.

Norgaard found that Norwegians have responded to this internal conflict by placing climate change outside their "norms of attention," which she defines as "the social rules that define what is or not acceptable to recognize or talk about." Thus, she says, people deliberately chose not to know too much in order to maintain their cultural identity as responsible citizens. "'Knowing' or 'not knowing,'" she says, "is itself a political act."

Like other academics who challenge climate change denial, Norgaard discovered for herself the mechanisms by which these politicized norms are policed and transgressors are punished. In 2012 a University of Oregon press release announced that that her research argues that cultural resistance must be "recognised and treated." The offhand comment—which did not even come from Norgaard—was eagerly distributed by Rush Limbaugh and other vocal climate deniers and led to a torrent of Internet abuse and pack bullying. A modest and diligent young academic now has her face permanently scattered across the Internet, photoshopped with swastikas, posted onto nude models, and stamped with vicious schoolyard abuse. One video about Norgaard was so offensive that YouTube removed it.

Martin Bursík, the former environment minister of the Czech Republic, described to me a similar distinction between individual facts and social facts in his country. Former Czech president Václav Klaus denied climate science—the only head of state in the world to do so. This, and a deeply entrenched social silence about the pollution of the coal industry, helped create a virtual taboo on public discussion of the topic. There is, Bursík said, not a single person in the Czech government who would be prepared to speak at an event on climate change. Privately, 92 percent of Czechs regard climate change as a serious or very serious problem, but, as Bursík explained, forty years of communist dictatorship has taught people to be very aware of boundaries defining what it is permissible to know. He says that they have learned "not to talk too loudly because the neighbors can listen through the wall."

In this context, it is not surprising that there are multiple similarities between the mechanisms of climate change denial and the socially constructed silence found around human rights abuses.

The late Stanley Cohen, a sociologist at the London School of Economics, drew strongly on his own experience as a Jewish South African to document the processes by which entire societies avoid dealing with collective human rights abuses. He highlighted the difference between ignorance (not knowing), denial (the refusal to know), and disavowal (the active choice not to notice). Of the latter, which he argued applied especially well to climate change, he wrote, "We are vaguely aware of choosing not to look at the facts, but not quite conscious of just what it is we are evading." So, Cohen pointed out, we notice that our neighbors disappeared in the night, or that cattle trucks are heading east full of people but returning empty, but somehow we also know that we should not talk about it.

Such comparisons are useful for the light they throw on human behavior, but we need to be careful of applying such historical experience too literally to climate change. There were real personal dangers in challenging the Gestapo or NKVD. Even within the Tea Party, the worst that would come from acknowledging climate change would be a degree of social estrangement.

It is, though, revealing that human rights organizations, which should be alert to the processes of social disavowal, have been so slow to acknowledge climate change, especially given that security analysts regard it to be a key driver of future conflict and forced migration. In 2006, the year that *An Inconvenient Truth* came out, I found that Amnesty International had not one mention of climate change on its website. It consistently received less than five mentions on the websites of many other leading progressive organizations—Physicians for Human Rights, the AFL-CIO, Oxfam America, CARE, World Vision, and the YWCA among them.

By way of comparison, I searched for two control terms that had no reason at all for being on these groups' websites: "donkey" and "ice cream." On each site, these irrelevant terms appeared far more often. Human Rights Watch mentioned donkeys four times more often than climate change. Refugees International mentioned ice cream nearly eight times more.

My interviews with decision makers in these organizations confirmed that they had *deliberately* framed climate change as being outside their norms of attention because they were unsure how they could intervene, had no capacity for uncertain emerging problems, and had decided that it was an environmental problem and therefore outside their mission. Only now that they have begun, belatedly, to make their voices heard, is climate change beginning to be recognized as a leading social rights issue.

Formal science is also subject to its own norms of attention about what it seeks to know or not know. A growing field of study is the "sociology of ignorance," which explores the field of "non-knowledge"—information that is deliberately not acquired because it is considered too sensitive, dangerous, or taboo to produce.

In-depth interviews by Joanna Kempner at Rutgers University found that scientists were committed to a heroic narrative of their pursuit of knowledge. "Our job," one said, "is to explore truth, not determining whether that truth is dangerous or that truth is unpleasant." However, while they welcomed controversy in their fields, they all wanted to avoid public controversy, which would take up their time and potentially threaten their funding streams. And who would not want to avoid the death threats or vicious abuse dished out to Kari Norgaard or Michael Mann? Scientists seek, in their own jargon to "lunatic-proof" their lives and to protect their families, and inevitably, this will determine what they study.

Ignorance is also generated by the academic process itself. As academics become ever more specialized and bury ever deeper into their silos, it becomes harder for them to acquire more generalized knowledge. There is a knowledge-ignorance paradox that the ever-increasing growth in knowledge leads to a simultaneous growth in what is not known.

By this reckoning, there is a great deal that is not known about climate change. As noted throughout this book, the people who take ownership of climate change consistently attempt to shape it in their own image, defined by their own discipline. Climate change, though, requires a much more creative and flexible approach that also considers what is not known or not said. It is not surprising that the vast majority of the specialists I have interviewed for this book are creative pioneers who resist easy categories and explore the boundaries between disciplines.

It was in this spirit of opening climate change to new perspectives and challenges—what is sometimes called *post-normal science*—that one of the largest British science research funding agencies invited me in 2012 to join its peer review board. I assumed it would be a chance to learn something about the scientific process. It turned out to be a far more interesting object lesson in socially negotiated silence.

The proposals we considered were mostly concerned with the impacts of climate change. Mostly. But interspersed throughout the backbreaking stack of photocopies were cuckoos: elegantly worded funding requests from oil companies asking for geologists and earth scientists to assist in the development of new oil and gas fields.

When I raised a question about the guidelines for funding oil compa-nies, it was met with utter silence. No one said anything. It all felt very uncomfortable, but I took a gulp and pushed on regardless—because, I said, "if we were in the Medical Research Council and had received a research proposal from a tobacco company, I think we would at least discuss it."

None of the imposing senior professors on the panel could answer the question. The proposal, they said, was well prepared. That was all. At this, the chairman brought down his guillotine. "We're not going to close down oil companies, are we?" he said, sighing. The proposal went through. As did every subsequent oil proposal for which the chair, with increasing grumpiness, repeated that "Mr. Marshall's objections are noted," without noting anything.

Eviatar Zerubavel laughs when I tell him this story—this was a text-book meta-silence. It was as if I had said nothing at all. It was all polite enough. We made empty small talk over the lunchtime sandwiches, again over the tea and biscuits, said our farewells. I have not been invited back.

Politicians and the media also have internal cultures that establish what can or cannot be recognized. The policy specialist Joseph P. Overton argued that there is a "window" that swings from left to right and defines what is politically possible to say or do. Overton argued that if politicians favor a policy that lies outside that window, they need to ensure that the window shifts to accommodate it—for example, by encouraging outside pressure or, as Naomi Klein argues in her book *The Shock Doctrine*, by enabling the emergence of social and economic crises that can then be used to justify radical measures.

Similar things happen in the climate change discourse. Extreme events, such as Hurricane Sandy, Hurricane Katrina, and Typhoon Haiyan, can shift the window to favor a political response, just as, the climate scientist Michael Mann argues, the leaking of scientific e-mails in 2009 combined with a cold winter swung the window in the direction of denial arguments and then silence.

The result was that politicians and campaigners increasingly stopped talking about climate change at all. Barack Obama made no major policy speech on climate change during his first term and, for the first time in twenty-four years, it was not mentioned once in the debates for the 2012 presidential election. John Kerry, to his credit, made an impassioned floor speech in the U.S. Senate denouncing the "conspiracy of silence . . . a silence that empowers misinformation and mythology to grow where science and truth should prevail."

At a state level, Republican legislatures then began to systematically remove all mention of climate change from policy. In North Carolina, state lawmakers passed a bill that forbade the use of any climate models for predicting future sea levels. In Texas, scientists had to mount a vocal revolt when officials attempted to purge all mention of climate change from their report on the environment of Galveston Bay. In 2013 nine states failed to mention climate change at all in their State Hazard Plans.

In other Republican states, planners have been allowed to develop long-term measures to ▉▉▉▉▉▉▉▉ impacts on the understanding that ▉▉▉▉▉▉▉ itself is never actually recognized or mentioned. And so the bizarre situation has arisen that Florida's and Arizona's populist leaders, who denounce ▉▉▉▉ science (Arizona governor Jan Brewer once punched a reporter for having the temerity to ask her whether she "believed in ▉▉▉▉▉▉▉"), are mandating their state, city, and county authorities to incorporate the latest ▉▉▉▉ models of drought and sea level rise into their long-term planning. Like Harry Potter, they have been actively preparing for a threat that cannot be named.

In March 2009, as momentum was building for a national climate bill, the White House distributed a memo to the leaders of U.S. environmental organizations demanding that they should not use the phrase "climate change" in regard to the bill and instead focus on "green jobs" and "energy independence." The bill itself was called the American Clean Energy and Security Act. Bill McKibben, alone among those present, stood up and protested. "This is going to come back and haunt us," he said.

Many environmental organizations concluded that their best chance of getting any political movement at all was by expunging all mention of climate change. Betsy Taylor, a specialist in environmental communications, complained that 2010 was the year when "it became a mantra inside big environmental groups. Talking about climate change is toxic. Some still don't use the 'C' word."

The C-word, indeed! In the year that the biggest-selling music hit, lauded at the Grammys, was rapper Cee Lo's bouncy ditty "Fuck You," climate change could only be referred to as *the C-Word*.

The removal of climate change from the political discourse in turn influenced the media, or more precisely, the editorial policy defining what areas may or may not be explored by journalists. David Fogarty, the former Reuters climate change correspondent in Asia, said that getting a climate story published became "a lottery" with editors "agonizing, asking a million questions, and too frightened to take a decision." In developing countries, journalists reported similar frustration in getting climate change stories past their editors.

In 2010 the *New York Times*, the so-called newspaper of record that sets the editorial agenda for much of the U.S. news media, did not run a single lead item on climate change. Two years later only 10 percent of U.S. television coverage of the unprecedented heat waves made any mention of the issue. The media silence is, apparently, a matter of policy rather than circumstance.

There is no simple model for socially constructed silence but rather another circulation system of complex feedbacks. Climate change finds no foothold in the conversations between workmates, neighbors, or even friends and family. It is not mentioned in the focus groups that define electoral messaging. It is polluted with cultural values. It becomes a toxic C-word for politicians and communicators. It is largely ignored by the media.

Each silence appears to be built on the other silences, but they have a common basis in the need to avoid anxiety and defend ourselves. From a psychoanalytic perspective, denial and anxiety are closely linked. Things that cannot be assimilated are repressed. As Stanley Cohen wrote about human rights abuses, "Without being told what to think about (or what not to think about), and without being punished for 'knowing' the wrong things, societies arrive at unwritten agreements about what can be publically remembered and acknowledged."

Of course this may yet change. The Overton window appears to be swinging back, pushed along by highly salient weather events. Action against climate change may not yet be a safe topic for a barroom conversation in Tulsa, but it is appearing again in the jokes of the late-night talk shows. The processes that define the norms of attention contain powerful feedbacks that can amplify change as well as suppress it. Over my lifetime, there have been remarkable (and hopefully unstoppable) shifts in public attitudes to race, homosexuality, child abuse, and disability. However, none of these occurred without a prolonged struggle by dedicated social movements, often with a central tactic of confronting a socially constructed silence. The lessons of history show that this is winnable, but it could be a long struggle.

18

The Non-Perfect Non-Storm

Why We Think That Climate Change Is Impossibly Difficult

PSYCHOLOGISTS WORKING IN THE FIELD of decision making often describe climate change as the perfect problem—so perfect, in fact, that one could easily conclude that we don't stand a chance in hell.

Tony Leiserowitz, of the Yale Project on Climate Change Communication, says, "You almost couldn't design a problem that is a worse fit with our underlying psychology." Daniel Gilbert says that "it really has everything going against it. A psychologist could barely dream up a better scenario for paralysis." And Daniel Kahneman is, of course, "deeply pessimistic."

As I continued to speak to people working in other disciplines, I found that, curiously, climate change happened to be the perfect problem from their perspectives too. Economists like Lord Stern describe it as the "perfect market failure." The moral philosopher Stephen Gardiner describes it as the "perfect moral storm."

This view was writ large in a major conference held at Yale University in 2005, which concluded that climate change is "almost perfectly designed to test the limits of any modern society's capacity for response— one might even call it the 'perfect problem' for its uniquely daunting confluence of forces."

So, is climate change really a perfect problem from all of these perspectives? Or does it just seem that way because the narratives that are constructed around it embody, so perfectly, the interests of the people who shaped them? This is an important question because defining climate change as the "perfect problem" triggers frames of powerlessness and hopelessness that feed denial.

In terms of proximity (one of the greatest cognitive obstacles), climate change is a mid-range problem. There are always threats that are even more distant in time and space. Down the road from me in Mid Wales is the Near Earth Objects Information Centre, which tracks extraterrestrial objects that might collide with Earth. It is a small, plain building that looks rather like a brick retirement bungalow. This is not so surprising, because the center is mostly funded out of the army pension of its director, Major Jay Tate. He and his doughty team of volunteers seem to be unable to get any decent funding for a threat that has no corporate interests (as the environmentalist Bill McKibben likes to say, no one makes fifty-five billion dollars in the meteorite industry) and that is so distant, so uncertain, and to many people so unlikely, that very few people take it seriously.

Except, strangely, climate change deniers. One of the center's strongest supporters is Benny Peiser, the director of leading British denial think tank the Global Warming Policy Foundation. Ten years ago, Peiser, who warns constantly about the exaggerated risk of climate change, was alerting anyone who would listen that a half-mile-wide asteroid, named NT7, was heading straight for Earth. In recognition of his contribution to asteroid worry, the International Astronomical Union named a six-mile-wide asteroid in his honor: 7107 Peiser is officially listed on NASA's website. Peiser's own website, meanwhile, routinely savages NASA's climate scientists.

Another enthusiastic supporter of the center is U.S. Representative Dana Rohrabacher, a Republican from California, who has been working hard to convene a House hearing on protection from "near Earth objects," which he describes as a "real and tangible danger." Rohrabacher is, however, adamant that we do not need any protection from climate change, which, he says, is categorically *not* caused by carbon dioxide. He has even suggested facetiously that prehistoric climate variation was caused by "dinosaur flatulence."

Whatever they say, climate deniers are clearly not rejecting climate science because of scientific uncertainty or the exaggeration of the threat. One is rather inclined to think that their entire pool of worry has been displaced by large rocks.

Nor is action to slow climate change impossibly costly. We may be reluctant to change our way of life, but everyone can remember what it was like to live in a society with lower emissions, and we all know that it was not that bad—in some ways it was better. By all of the indices, happiness in developed nations peaked in the early 1970s, when Americans drove 60 percent less and flew 80 percent less. Was it so bad? The really important things: family, friends, community, faith, joy, excitement, laughter, passion, and beauty could, if anything, be enhanced in a low-carbon society. And sex is, beyond any argument, entirely carbon neutral.

So addressing climate change may be challenging, but it is not the perfect cognitive challenge by a long stretch. Problems that are without any precedent, or are reported by entirely unreliable sources, or affect our closest relationships—now those are the hard ones. The 1956 sci-fi film *Invasion of the Body Snatchers* ends with the disheveled Dr. Miles Bennell desperately trying to persuade the FBI that his friends and children had been taken over by aliens who hid themselves in giant sea pods. Now it really would be hard to get anyone to believe in that.

Then again, strangely, some people do. In 2012 David Icke, a new age guru, managed to fill the largest football stadium in the UK for an eleven-hour monologue on the takeover of shape-shifting reptilians from the constellation Draco. Apparently these "Reptoids," who now rule the world, have taken human form and include the Queen, Al Gore, and the entire Bush family. Icke describes climate change as a "monumental scam," showing, once again, that people can believe just about anything if it lines up with their worldview.

Taking a step back, climate change has many aspects that are considerably less daunting than they might otherwise have been. These are not reasons to be cheerful but perhaps reasons to be somewhat less despondent.

For example, it is extremely fortunate that climate change is occurring now, during the longest prolonged peace in the developed world since the emergence of the modern nation state, and at a time when we have the combination of technology, wealth, education, and international cooperation that might be able to respond to it. This is not perfect timing, but it is just about as good as it could be. What is more, the countries causing climate change will also be impacted by it. While this clearly has its own misfortune, one positive result is the increased likelihood of action by these countries. If that self-interest did not exist—if, say, the extreme

impacts were concentrated entirely in Africa—I fear that developed countries would do absolutely nothing.

So climate change is very difficult, but not *perfectly* difficult. In theory
we can deal with this—it is all a matter of degree. Humans are smart,
but are we smart enough? We are cooperative, but are we cooperative
enough? Those people who understand and get passionate about this
issue are numerous, but are they numerous enough? We have a little
time, but is it enough?

If there are any grounds at all for regarding climate change as a
"perfect" cognitive challenge, it is not because of its specific qualities but
because it is so *multivalent*—that is to say, it is so open to multiple meanings and interpretations. It provides us with none of the defining qualities
that would give it a clear identity—no deadlines, no geographic location,
no single cause, solution, or enemy. Our brains, constantly scanning for
the cues that we need to process and categorize information, find none,
and we are left grasping at air. But we still need these cues—we cannot
deal with it otherwise—and so we create and impose our own. This is a
dangerous situation, leaving climate change wide open for miscategori-
zation, confirmation bias, and the opportunity for us to "believe what we
want to believe."

This is the reason, when asked why we are taking so little action on
climate change, that everyone seems to shape the problem in his or her
own image. Climate scientists say that people don't understand the
science. Environmental campaigners say that the political process is
corrupted by oil companies. Oil companies say that the political process
is corrupted by environmental campaigners. Mark Berliner, a professor
of statistics at Ohio State University, says that our failure comes from our
"aversion to statistical thinking." And communications specialists such
as myself say—lo and behold—that the main reason why people have not
responded to this threat has been because of failed communications. If
climate change really is the "elephant in the room," it is a pitch-black
room, and, like the blind men in the ancient fable, we are all feeling
different parts of it and drawing our own, culturally biased conclusions
about what they might be.

But the ambiguity of climate change extends even wider and threatens
to infect those things that already seemed most safe, secure, and familiar.
It suggests that the way of life that we associate with our comfort and the
protection of our families is now a menace: that our well-meaning actions

might hurt the ones we love, that gases we have believed to be benign are now poisonous, that our familiar environment is becoming dangerous and uncertain.

Drawing on folktales and horror stories, Sigmund Freud, the father of psychoanalysis, recognized the destabilizing psychological impact of something that seems to be almost familiar, yet is not. He called this *das Unheimliche*, usually translated into English as the "uncanny condition." Climate change is inherently uncanny: Weather conditions, and the high-carbon lifestyles that are changing them, are extremely familiar and yet have now been given a new menace and uncertainty.

So we have a dangerous combination. Climate change is exception-ally *multivalent*, enabling a limitless range of self-serving interpretation. And it is *uncanny*, creating a discomfort and unease that we seek to resolve by framing it in ways that give it a familiar shape and form. These two factors combine, to add a third term to the mix, to make it an exceptionally *wicked problem*.

The concept of a "wicked problem" was first formulated in 1973 by Horst Rittel and Melvin Webber, urban planners at U.C. Berkeley. Originally they applied the concept to policy planning, though in recent years it has gained much wider currency because it fits so well with intractable global issues such as terrorism, financial crises, and, of course, climate change, which has often been called the "ultimate" wicked problem.

Simple problems, what Rittel and Webber called "tame problems," have defined causes, objectives, and outputs. Wicked problems, though, are multifaceted in every respect—they are incomplete, contradictory, and constantly changing. Tame problems may be very complicated, but wicked problems are *complex*. As a result, there is no point at which one has enough information to make decisions. Instead, wicked problems demand a continuous process of evaluation and redefinition.

Really, though, there is a no definitive definition of a wicked problem other than to say (take a deep breath here) that, by definition, it defies having a clear definition because it keeps evolving according to the solu-tions we evolve to solve it. You can't learn about a wicked problem without trying solutions, but every solution you try creates new consequences and new wicked problems.

A problem becomes wicked when it is a symptom of a large chain of adjacent issues, with multiple partners, whose understanding of the problem is also dependent on their own different ideas for solving it. Thus, every different definition of the problem generates a different set of solutions. And every set of solutions creates a different definition of the problem.

We can define climate change as an economic problem, a technological problem, a moral problem, a human rights problem, an energy problem, a social justice problem, a land use problem, a governance problem, an ideological battle between left and right worldviews, or a lack of respect for God's creation. Each approach will generate different responses, different ways to share the costs, and, especially, different language with which to justify action. Or inaction, because some people will refuse to accept that climate change is a problem at all. This is yet another quality that it shares with other wicked problems.

For Rittel and Webber, the fundamental rule for handling wicked problems is that they must *not* be treated like tame problems. Tame problems can be solved by a series of distinct steps: First, understand the problems, then gather information, then pull that information together, and then work out and apply solutions. For wicked problems, however, this type of scheme does not work. They argue that one cannot understand a wicked problem without knowing about its context, one cannot search for information without knowing the solution, one cannot first understand, then solve.

Climate change refuses to fit any structure of cause and effect because it is never clear whether one is looking at the actual cause, or a cause created by the way we have chosen to define the problem. Do we find climate change hard to accept because it lies in the future, or have we chosen to place it in the future to make it hard to accept? Is climate change really uncertain, or do we just tell it that way? Are the solutions too challenging, or do we just describe them as such?

So calling climate change the perfect problem is just another attempt to define it and tame it. Is it a perfect problem, a non-perfect problem, a non-problem? Or is it none of these things and just (choosing a word that best expresses our own biased interpretation) extremely, irrevocably, singularly, horribly_____?

IT IS AGAIN TIME TO recap what has been said so far.

I have shown that the processes of attention are fundamental to our thinking. We are "wired" to scan incoming information for the cues that tell us whether we should pay attention to it and how we can categorize it.

Many of these processes are innate and intuitive, formed early in our long evolutionary history. The issues with greatest salience—that demand our attention—are those that are here, now, and contain a clear visible threat from an identifiable enemy. I argued in the earlier chapters that the social cues provided by the people around us also compel us to pay attention and I have shown that climate change is often subject to a socially constructed silence that strips it of these cues.

Without salience or social cues climate change sits outside the analytic frame that we apply to make sense of the world around us. Rather than actively attending it, we actively disattend it, keeping it permanently on the edge of our "pool" of worry.

However, I argued, this is not because climate change is innately unthreatening. Clearly it poses a very serious threat and our rational brain can fully appreciate this. The problem is that the signals it supplies to our other emotional brain are far too ambiguous to galvanize us into action. Climate change is here and now, but it is also there and then. It does have causes and impacts, but these are widely distributed. It refuses to fit into any single category and, as a result, fits into none. It is exceptionally *multivalent* and, as a result, it invites us to apply our confirmation bias and to "believe what we want to believe."

For this reason the critical means by which we make sense of the issue is the way that we talk about it—and in particular the stories we build around it. And, as I will show, it is these socially constructed stories, not climate change itself, that people choose to accept, deny, or ignore. Climate change may not be the perfect problem, but it does not generate the perfect story either, and this, as I will explain, may be its biggest problem of all.

19

Cockroach Tours

How Museums Struggle to Tell the Climate Story

THERE IS A SCHOOL CLASS inside the Hall of Human Origins at the Smithsonian Institution's National Museum of Natural History when I visit, and a young girl is reading out the panels to her friend. Some of the language is a bit tricky, but she is giving it her best. "This ex-hibit shows how the charact-character-istics that make us human evolved over six million years as our an-cestors struggled to survive during times of dramatic climate change." She gives me a big grin and asks, "How did I do?"

The hall has molds of footprints and skulls in well-designed if some-what staid displays to show evidence that climate change has been a driver in human evolution. This theory is generally supported by paleon-tologists, though it is a simplification and excludes the many other factors that contributed to brain development, such as sexual selection, coopera-tion in social groups, hunting, and cooking. Museums like to tell a clear and digestible story, and so have kept this one simple.

Actually, though, the Hall of Human Origins is telling two stories: One is about evolution. The other is about climate change. The first thing one sees on entering is a ten-foot-wide video screen on which the phrase "Extreme Climate Shifts" is repeated over and over again on top of images of the cracked earth, deserts, and melting ice sheets—all familiar images from the B-roll of environmental documentaries. Climate change carries powerful and resonant frames and these are triggered by the language and images in every display inside the hall.

Yet, while toying with these associations, there is a strange absence of discussion about climate change as a threat in our own times. A single panel, tucked over to one side, tells us that carbon dioxide levels are increasing and that these are "associated with" a warmer planet and sea level rise. Alongside is an interactive video game that weaves this potentially distracting information back into the central theme of the gallery. It invites visitors to consider what useful new features we might evolve to cope with this greenhouse future. "Imagine all the land is underwater," it says. "Would you have big webbed feet like a duck or long stilt legs like a flamingo?" "The temperature is really hot—do we have a tall narrow body like a giraffe or more sweat glands?" Each time you press a button, the cartoon figure alongside develops a new body shape or a shower of droplets pours from his armpits. The schoolkids are shrieking with laughter as they put together the freakiest possible future man.

This is not unlike the deliberately provocative proposal by the philosopher S. Matthew Liao of New York University that we should genetically engineer humans to reduce emissions. We could have cat eyes so we need less lighting. We could have a choice between two medium-size children or three small-size children in order to save energy. "Examining intuitively absurd or apparently drastic ideas can be an important learning experience," Liao says.

If so, the children in the Smithsonian are learning nothing about climate change. There is no information about the issue, its causes, or its solutions. But they *are* learning some powerful frames. The narrative that the Hall of Human Origins promotes to the million plus people who visit every year is that the climate has always changed, that we have always coped with these challenges, and that adapting to them is what has made us strong and smart.

It is an argument that would chime very happily with the fossil fuel lobbyists and professional contrarians in their K Street offices just a few blocks from the museum, most of whom have been directly funded by those nefarious oil billionaire Koch Brothers. The Kochs are men of many interests who like to spread their largesse around, including—oh, didn't I mention this?—twenty million dollars for the *David H. Koch* Hall of Human Origins.

Talking later with Kirk Johnson, the director of the Smithsonian Natural History Museum, on the well-stuffed sofa in his vast corner office on the fourth floor, I am intrigued to know why the Smithsonian, America's most respected scientific body, took funding from America's most notorious climate change denier to host a permanent display—one that now carries his name—portraying climate change as a natural cycle and positive challenge that we will mutate to survive.

Johnson is up for the challenge. He understands the battlefield of climate communications well. His own initiation came ten years earlier, when he made a keynote presentation on climate science to the great and good of the Albertan oil industry. "They were really mad! I was in the full blast of two thousand really unhappy people. And all the time I was thinking, 'Wow, that's interesting.'"

Maybe I am too easily charmed by Johnson, who is smart and entertaining, but he persuades me that there was never a deliberate strategy to misinform: This was a paleontological display on the effects of *natural* climate change, which just happened to stray, rather naively, into the highly politicized issue of *anthropogenic* climate change. Johnson, who has a background in paleoclimatology, freely admits that the two have been confused in the display and that the cold-to-cool changes described in the exhibition are entirely unlike the devastating cool-to-hot changes we are facing with human climate change.

According to Johnson, there was also a great match between a curator with a specific vision and a large donor who is so passionate about paleontology that he makes regular visits to research sites in the Rift Valley. The enemy narrative would love to read Koch's support as another cynical exercise in public misinformation, but reality is more complex and interesting than that.

Nonetheless, I doubt that David Koch is the "hands off, here's the money, see you later" funder described by Johnson. No one wants to antagonize their largest donor, and I would happily bet that the museum held back on any temptation it might have had to make the hall that bears his name into a showcase for the latest climate science.

Which raises a more interesting question: Why is a single lackluster panel in the corner of a paleontology gallery the only permanent display, across the entire Smithsonian museum complex, of the most important science-driven issue of our times? The real problem seems to be that people at the Smithsonian, like everyone else who works on climate change, are struggling to find ways of talking about it that are interesting, engaging, and truthful to the science yet able to navigate the politics. And, it would seem, they are not doing a great job of it.

Johnson says that he has never seen any museum, including his own, address climate change effectively. There is, he feels, a consistent mismatch between the subject matter and the audience, which tends to be schoolchildren and families on holiday. You get people for only about two hours, with breaks for food or the bathroom—and so you don't have a chance to educate. Maybe, he says, you can achieve "an inoculation of curiosity" that might trigger people to get more interested. But you have to find techniques that are interesting, startling, and fun to get the message out.

So, surprisingly, museums face exactly the same problems as any artist or entertainer: how to be faithful to the science, honest, and independent while avoiding the poisoned narratives of the partisan debates.

This was the challenge facing Professor Chris Rapley when he was appointed the new director of the Science Museum in London in 2007 and set up a team to design a six-million-dollar gallery about climate change. Rapley, a climate scientist himself, was the former head of the British Antarctic Survey. He feared that critics would accuse him of being an activist and kept a careful distance from the project. In summer 2009, though, with the permanent display still in the design stage and the Copenhagen Climate Change Conference just months away, he personally oversaw a temporary display called "Prove It!—All the Evidence You Need to Believe in Climate Change." The experience, he says, left him "badly burned and shaken."

As its name suggests, the display took the position that the science was beyond doubt and that the museum, with the help of disaster images and apocalyptic framing, could convince any remaining Doubting Thomases. It invited visitors to endorse a message of encouragement to governments at the Copenhagen climate conference to "prove they're serious by negotiating a strong, effective, fair deal." To Rapley's horror, deniers got into the online system with automatic voting software and sent in thousands of opposing votes effectively saying, in his words, "This is all bollocks and you should not support it." Rapley looks crestfallen: "I still blame myself," he tells me.

The museum trustees went into a meltdown. Rapley stepped in to drastically overhaul the flagship project, as yet still on the launch pad. In interviews he promised the right-wing press that the permanent gallery would not state a position on whether or not climate change is real and driven by humans, so as not to "alienate any people who want to be part of the discussion."

And so the Atmosphere Gallery—a name that embodies this neutrality—tries valiantly to meet its impossible brief. There is certainly a lot of atmosphere in Atmosphere. It's a cavernous blue-black space with five themed islands. It aims to be an "immersive environment," but feels rather like IKEA—once inside and disorientated, you trundle between spotlit zones, filling your brain with things that capture your fancy.

In the fashion of modern museums, the displays have been converted into computer game consoles. When I visit, one schoolgirl is immersed in a game involving loading a virtual gun with insulation and firing it at a house. She seems to have no idea of what this means but is enjoying firing a gun at things. At the next island, two boys tell me with teenage confidence that climate change does not exist and is a natural cycle. Their teachers, flirting in the nightclub gloom, tell me that they love the gallery because "it has lots of things to occupy the kids for half an hour before lunch. It gives us a much-welcome break."

By the critical measure of visitor numbers—known as "footfall" in museum jargon—the gallery has been an unexpected success, with more than a million visitors in its first fourteen months, four times above its target. According to Rapley, the exit polls suggest that it has fulfilled its mission to inform the concerned and engage the unconcerned. Overall, people said that "the tone of voice is what I would expect from the Science Museum."

But, I wonder, is it the tone of voice that we would expect for a crisis that threatens our survival? Nothing in the gallery suggests that this is a disaster, a historical turning point, an opportunity or, indeed, that anyone cares much at all. In fact, it seems that in its desire to avoid the contested narrative, the museum has largely abandoned narrative altogether: except, maybe, to show that scientists can do clever things with fancy instruments packaged with a vague technological optimism that, in Rapley's words, "human ingenuity can lead to a better future."

Rapley accepts that "possibly we overreacted and went too far into the neutral voice in the exhibit." The original idea was to bundle up the more contentious climate change parts into a parallel education program. This dissenting voice now survives largely as the "cockroach tours," in which people dress up in bug costumes and learn "how strange humans are that they don't confront these huge issues when they should do."

Maybe these quizzical cockroaches might like to linger a while at the entrance to the Atmosphere gallery and ponder the strange plaque honoring its principal donor, Shell Oil. On the plaque, Shell explains that "all of us need energy to develop and grow. That's why, at Shell, we are working hard to build a new energy system while supporting a deeper understanding of climate science."

Rapley is very defensive of the decision to take funding from Shell, saying that "demonizing the corporate world is a route to nowhere. It is much better to enter a dialogue with them." Certainly Rapley entered a very positive dialogue with the chairman of Shell, James Smith, who, he says, "was one of the most thoughtful people you can imagine and takes this issue very seriously." He is less enthusiastic about dialogue with the radical climate change campaign group Rising Tide, whose members invaded the museum with noisy protests and banner drops. Anyway, he insists, Shell "was under a very strict and legally binding contract" that it had no editorial influence over the exhibition.

But then why would Shell need to influence an exhibition that was so determined to say nothing that could challenge its interests? Shell received what they needed—a showcase for their global brand and own corporate story line—that they are good guys who produce energy, find positive environmental solutions, and help solve the climate crisis. Shell's narratives are very much better-written and better-funded than any climate change education program. It is a measure of their success that people not only accept them but cannot even see that they exist.

And so a museum with a proud scientific history, led by a dedicated and self-reflective climate scientist, ended up with a nicely lit atmospheric crèche sponsored by a company whose entire business model is, by necessity, based on making climate change worse.

Now that's an interesting story.

20

Tell Me a Story

Why Lies Can Be So Appealing

STORIES ARE THE MEANS BY which we humans make sense of our world, learn our values, form our beliefs, and give shape to our thoughts, dreams, hopes, and fears. Stories are everywhere; in myths, fables, epics, histories, tragedies, comedies, paintings, dances, stained-glass windows, films, social histories, fairy tales, novels, science schemata, comic strips, conversations, and journal articles. Before we can even read and write, we have learned more than three hundred stories.

In his book *The Storytelling Animal*, Jonathan Gottschall says, "We are, as a species, addicted to story. Story is for a human as water is for a fish." The author Philip Pullman, who has been among the handful of writers struggling to build stories around climate change, says, "After nourishment, shelter, and companionship, stories are the thing we need most in the world."

Stories perform a fundamental cognitive function: They are the means by which the emotional brain makes sense of the information collected by the rational brain. People may hold information in the form of data and figures, but their beliefs about it are held entirely in the form of stories. Stories are the essence of climate change as a wicked problem— where the problem is shaped by the very process of explaining it.

Professor Walter Fisher, a theorist in communications theory at the University of Southern California, has argued that when non-experts make sense of complex technical issues, they make their decisions based on the quality of the *story*—what he calls its "narrative fidelity"—rather than the quality of the *information* it contains. Fisher says that when we encounter a new issue, we ask: "Does it hang together? Does it contain a linear sequence of events from past to future? Do the characters behave as we would expect them to behave, with clear and understandable goals and motives? Does it match our own beliefs and values?"

So, stories strip facts away, seeking what is most narratively satisfying, not what's most important or truthful. They can be regarded as facts in their own right, gaining weight through repetition and social proof as part of a social norm—as happens in rumors.

People will maintain their belief in an engaging story even if they are told that it is a fiction. In one psychology experiment, people were invited to read stories that, they were clearly warned, were false. Later on, when they were given a general knowledge quiz, this incorrect information then reappeared in people's answers. They had internalized this information so effectively that some people could not remember that it had come from the stories they had first heard a few hours earlier.

For twenty-five years, psychologists have been repeating variations of another storytelling experiment. Participants are told the story of a warehouse fire in the style of live, rolling news coverage. First they hear of toxic smoke, then explosions, and then they are told that it may have been caused by gas cylinders and oil paints that were negligently stored in a closet.

The final story is so complete that many people resolutely refuse to accept any further variation that might weaken it. If they are subsequently told that there was no gas or paint in the closet, the repetition of the phrase leads some people to become even more convinced that gas and paint *were* responsible. Only if they are supplied with an even more compelling replacement story—for example, that arson materials were found in the closet—will they abandon the original version.

This experiment provides strong clues about what makes a compelling narrative—cause, effect, a perpetrator, and a motive (ideally one that is consistent with our assumptions about how we believe they might act). The most compelling narratives in climate change have this structure: Governments (perpetrators) justify carbon taxes (effect) in order to extend their control over our lives (motive). Right-wing oil billionaires (perpetrators) fund climate change denial (effect) to increase their wealth (motive).

This is why it is extremely hard for a deeply unengaging narrative based in fact to compete with a compelling narrative based in falsehood. "The balance of evidence leads many scientists to suggest that our emissions may be damaging the climate" is, unfortunately, less inherently compelling than "rogue scientists are conspiring to fake evidence in order to secure larger research grants."

As communications guru Frank Luntz advised Republican candidates on messaging climate change, "a compelling story, even if factually inaccurate, can be more emotionally compelling than a dry recitation of the truth." Luntz claims to have personally sat in more than two hundred focus groups, from which he has extracted his own rules for what constitutes a compelling story: simplicity, brevity, credibility, comprehension, consistency, repetition, repetition, and repetition.

Patrick Reinsborough, executive director of the Center for Story-based Strategy, suggests some more ground rules for a compelling story: simplicity of cause and effect, a focus on individuals or distinctly defined groups, and a positive outcome. A perfect example of a story that combines these qualities was the rescue in August 2010 of thirty-three Chilean miners who had been trapped for two months down a San José gold mine. A staggering one billion people watched their rescue broadcast live on global television.

The ability to identify directly with the victims is a consistent component of compelling stories. As Joseph Stalin put it, "The death of a single Russian soldier is a tragedy, but a million deaths is a statistic." Not that Stalin seemed overly concerned either way, but he knew good propaganda when he saw it. In experiments, people overrule factual evidence in favor of vivid personal descriptions and donate generously to feed a single child whose name they know and photograph they can see, but give somewhat less to feed two children and only half as much to feed "millions of Africans."

Another way to identify the components of compelling stories is to look at existing stories and strip them down to their core. Few people have approached the task with such obsession as the journalist Christopher Booker, who spent thirty-four years condensing a thousand classic novels, films, plays, and operas down to seven basic plots. These, he says, all originate from the same great basic drama, in which a hero or heroine is constricted but ends up with "a final opening out into life, with everything at last resolved."

It is maybe no coincidence that Booker is also Britain's leading climate change denier. He was forced to pay a large libel settlement and publish a full retraction after telling an entirely fabricated story that the head of the Intergovernmental Panel on Climate Change, Rajendra Pachauri, made "millions of dollars" from his links with carbon trading companies. One critic said that Booker has "been playing to the gallery for so long that he can no longer distinguish between fact and fiction."

Much the same could be the late Michael Crichton, the only writer to have achieved a simultaneous number-one slot for a television program (*ER*), a film (*Jurassic Park*), and a book (*Disclosure*). Crichton drew on all the classic narrative devices in his 2004 eco-thriller *State of Fear*, in which eco-terrorists from the Environmental Liberation Front set out to trigger natural disasters in order to create mass panic about climate change and install a green dictatorship. It presents a compelling narrative with a focus on individuals, a struggle against an evil conspiracy, and a positive outcome. It is, unfortunately, by far the bestselling book yet written about climate change and includes dense technical appendices to prove that it is a concocted myth. *State of Fear* presents rather a curious paradox: that the truth is portrayed as a story within a story that is perceived, by many readers, as truth.

Not least by President George W. Bush, who spent an hour chatting with Crichton about the book in the Oval Office, after which, according to his chief of staff, they were in near-total agreement. The novel was then presented as "scientific" evidence into a U.S. Senate committee and Crichton gave briefings on climate change around the world at the invitation of the U.S. State Department.

The critical ingredient that has made *State of Fear* such powerful denialist propaganda is that Crichton perfectly understood the principle of narrative fidelity and set out to write a compelling story. It has pace, enemies, motives, and a comprehensible human-generated threat that could be defeated. Like all good myths, it ends with the punishment of the perpetrators and the restitution of social order. It is hard to think of any story that could be more different from the complex, multivalent, collective, and boundless reality of climate change.

21

Powerful Words

How the Words We Use Affect the Way We Feel

WORDS ARE POWERFUL. EVERY TIME a word is used, it brings into play a cluster of interlocking frames and associations. Sarah Palin's offhand phrase "death panels" generated such a powerful frame that it nearly derailed Obama's healthcare plan. Progressives play this game too, and the insertion of their own language, such as gay, same-sex, and African American, into their opponents' vocabulary is a lasting campaign achievement.

George Lakoff, the U.C. Berkeley cognitive linguist, argues that the goal of good communications is to use "the words that trigger your frames and inhibit your opponents' frames." Once words become engrained in common usage, they forever carry around their frames.

An example that Lakoff often gives of political reframing is the phrase "tax relief," which replaced "tax cuts" and was inserted, quite deliberately, into the early speeches of George W. Bush. The word "relief" activates the frame that taxation is an affliction, that the person who relieves that affliction is a hero.

As a result, "tax" has become a deeply problematic word for many conservatives. In one experiment, Republicans were five times more willing to pay a 2 percent climate change surcharge on an airline ticket when it was described as a "carbon offset" than when it was called a "carbon tax." They were asked to write down their thoughts as they considered their choice, and it was clear that the mere word "tax" had triggered a cascade of negative thoughts that had biased their entire decision making.

Knowing this, the lobbyists for a climate change bill in the U.S. Senate deliberately excised the word "tax" and replaced it with the more anodyne "fee on polluters." Then Fox News brought it right back into play with an online article opposing the bill that called it a "tax" thirty-four times.

Another recent framing battle emerged over the name "bituminous sands." The common term for these was "tar sands," until the Canadian oil industry became concerned about framing and decided to rebrand them as "oil sands." Campaigners, of course, preferred the old frame. The Canadian Broadcasting Corporation, forced to adjudicate, mandated the term "oil sands" in its reporting on the grounds that this is "more accurate because the substance refined from the extracted bitumen is oil." One campaigner wryly commented that on this basis we should call tomatoes "ketchup."

The most effective and deliberate linguistic framing occurred in November 2009 when the server of the Climate Research Unit of the University of East Anglia was hacked into and fragments from a thousand e-mails were pieced together by climate deniers into a grand narrative of conspiracy, manipulation, and suppression of dissent. They dully dubbed it Climategate (the rival term Climaquiddick never took off), and within a week this new word had appeared more than nine million times across the Internet.

This was hardly a moment of great linguistic originality, following, as it did, the previous manufactured scandals of Nannygate, Nipplegate, Grannygate, Flakegate, Tunagate, Biscuitgate, and Pastagate. But it was a textbook example of how to use framing to dominate a discourse with your own values. Two months later, I attended a communications conference at which every participant used the term, even a senior professor from the Climate Research Unit itself. Climate deniers keep hoping to replicate that initial success with Himalayagate, Amazongate, Glaciergate, and Hurricanegate.

The e-mail scandal shows the power of words and the frames they trigger. The main allegations revolved around a single phrase in the e-mails: "let's use Mike's trick to hide the decline." The decline can be readily explained (it is a decline in tree growth, not temperatures), but it was the use of the word "trick" that triggered the frame of deceit. "Trick" is being used here in a specific scientific usage to mean a clever and elegant solution, but was understood by the lay public in its more common meaning of fraud. The fact that "Mike" was the much-abused climate scientist Michael Mann added an identifiable and familiar enemy to the story.

This is just one of a number of dangerous false friends (words that sound the same but mean something different) that bedevil climate change. In general scientific terminology, uncertainty, theory, error, and manipulation all have far more precise meanings in science than in general use. All the terms in the core lexicon of climate change (anthropogenic, albedo, aerosols, radiation, emissions, greenhouse gases, mitigation, adaptation) emerge from within the scientific discourse without any consideration for their framing. The words *enhance* and *positive* suggest an increase or change to scientists when, to the general public, they suggest an improvement.

It is the U.S. scientist Wallace Broecker who has the dubious claim of having invented and then first used both of the core terms for this problem in a single 1975 article: "Climatic Change: Are We on the Brink of a Pronounced Global Warming?"

Having two terms is, in itself, a problem, generating confusion and a politicized battle to promote the term that each side assumes will serve its interests. In the late 1980s, the United States and Saudi Arabia lobbied during the world climate negotiations for the language of early resolutions to be changed from "global warming" to "climate change" on the assumption that this sounded less emotive and, more important, had less connection to the burning of fossil fuels.

In an internal memo to Republicans in 2003, communications consultant Frank Luntz argued that the term "climate change" sounds more moderate and controllable. As evidence, he cited one focus group participant saying that climate change "sounds like you're going from Pittsburgh to Fort Lauderdale." The Bush administration duly followed his advice, and President Bush adopted the term "climate change" in all subsequent speeches. Ironically, climate deniers now accuse environmentalists of seeking to suppress the phrase "global warming" because, they claim, temperatures are no longer increasing. Maybe they should all have waited for some better research: Testing conducted with greater rigor shows that Republicans are more likely to believe in "climate change" than in "global warming."

Environmental campaigners hate both terms and seek, intermittently, to introduce new phrases. Earth scientist James Lovelock for example, complains that global warming sounds like "a nice duvet on a cold winter's day" and advocates for the term "global heating." Other proposals have included "global weirding" and "global climate disruption," and Al Gore has contributed neologisms like "climate chaos," "climate crisis," and more recently, "dirty weather."

However, although neither "global warming" or "climate change" is ideal, neither is disastrously bad either, and "climate change" has a sufficiently bland emptiness that it allows people to fill it with their own meanings. Seth Godin, a communications specialist, wonders whether calling it "Atmosphere Cancer" or "Pollution Death" might not have garnered more concern. It's unlikely, since to anyone conservative the terms sound outrageously biased, and to anyone else they sound like heavy metal bands.

Language around climate change is constantly evolving as users experiment with new combinations. "Climate" becomes the single-word signifier for the wider problem (climate crisis/scientists/skeptics/ deniers), and the word "carbon" becomes the signifier for the emissions that cause it, such as the term "carbon neutral," which was chosen by the 2006 *New Oxford American Dictionary* as its "word of the year."

"Carbon" is a dead and emotionally meaningless word without aspiration or inspiration, so, like the element itself, this word is particularly prone to forming compounds with other words. A team at Nottingham University studied the growth and spread of thirty-four new "carbon" phrases in the media and Internet. It started in the 1990s with technical accounting terms (budget, market, credit, allowance, tax). By the early 2000s, it became associated with terms for personal consumption (friendly living, conscious lifestyle, diet), and more recently, it merged with more morally laden meanings (addiction, guilt, dictatorship, indulgence, crusade, morality).

It is unfortunate that the most common compounds of all, *high carbon* and *low carbon*, are used to differentiate lifestyles, economies, and technologies. "High" is a universal frame for status and power. We say high-class, high-end, high quality, high achievement. "Low" is a universal frame for inferiority and social failure. No matter how much you try to bend it, "high-carbon living" sounds intuitively like having champagne in a penthouse and "low-carbon living" sounds like drinking cold tea in a dank basement.

In 2007 the U.S. Supreme Court decided that carbon dioxide could be defined as a pollutant and subject to regulation by the Environmental Protection Agency. In all his speeches, President Barack Obama has duly promoted a new carbon compound, the term "carbon pollution."

This is part of a deliberate and intelligent attempt to reframe climate change in terms of health, purity, and progress. "Carbon pollution" triggers the deep frames of dirt, corruption, and illness; in contrast, "renewable energy" is promoted as a "clean" energy with associations of cleanliness, freshness, and bright sunlight—and by association, with health, life, and youth. This time, at least, all the narratives, frames, and metaphors line up perfectly, and this is supported by focus group and survey research showing bipartisan support for "clean energy" and action against pollution.

This does not mean that people will readily regard climate change as a form of pollution. Carbon dioxide is hardly ideal as a "pollutant." It is invisible, odorless, and harmless to human health. Generations of schoolchildren have been taught that it is a natural gas that is vital for life on Earth and that is directly linked with industrial progress and plant growth. Back in the 1950s, Harrison Brown, an atomic scientist who had worked on the Manhattan Project, claimed that we could double food production by increasing atmospheric carbon dioxide. His project, which was endorsed by Albert Einstein, proposed ringing the world with huge coal-fired "carbon-dioxide generators." It was an insane experiment that we seem to be determined to carry out to the letter.

So, will this attempt to reframe carbon as *pollution* be successful or make any difference? Possibly, but as Dan Kahan of the Cultural Cognition Project at Yale Law School would say, the primary pollution of the issue is from the social meaning that people create rather than the terms themselves. Frames amplify social meaning, not replace it. In any case, things often thrive with bizarrely inappropriate names. RadioShack? Craigslist? Sometimes you just have to work with what you have.

The other building block of narrative is metaphor. Through metaphors, we mobilize our most "available" previous experience to make sense of new information. Metaphors are usually culturally coded, and the choice of a metaphor is itself a social cue, drawing on a back catalog of associations and meanings.

While it is the case that climate change is unprecedented—no human action has altered the very climate itself—the way we think about it

involves precedents. As far our brains are concerned, there really is nothing new under the sun. On occasion, communicators just throw the
whole lot at it in the hope that something sticks. As one blog said recently:

> "The inhabitants of planet Earth are quietly conducting a gigantic experi-
> ment. We play Russian roulette with climate and no one knows what lies
> in the active chamber of the gun." If the Nazi's constructed gas chambers
> for millions of victims, ongoing climate change threatens to turn the
> entire Planet into an open oven on the strength of a Faustian Bargain.

Phew.

Everyone likes to talk about World War II—the most recent experience
of mass mobilization for a common purpose against a common enemy.
The cover of the April 2008 *Time* magazine special issue titled "How to
Win the War on Global Warming" was a parody of the most famous
photograph of the war, the raising of the flag on Iwo Jima, except that this
time the soldiers are raising a tree, not a flag.

The Second World War provides an inspiring model for cooperation
and shared sacrifice against a common threat. It lives on in climate
communications in quotations from Winston Churchill, Rosie the
Riveter, the home front, and the holocaust.

Deniers draw on the same wellspring of metaphor. Myron Ebell of the
Competitive Enterprise Institute, with typical inversionism, says that the
struggle over climate change is "a desperate last-ditch Battle of the Bulge
type effort by the forces of darkness." He is talking about good versus evil,
Allies versus Nazis, who, in this case, are standing in for environmentalists and their liberal allies.

Godwin's law, created by Mike Godwin in 1990, states that in any long
online discussion, regardless of its topic or scope, someone inevitably
makes a comparison to Hitler or the Nazis. No current issue is quite so
plagued, on both sides, by resurgent Nazis. Al Gore draws a parallel
between fighting global warming and fighting the Nazis. William Gray, a
skeptic professor at Colorado State University, says, "Gore believed in
global warming almost as much as Hitler believed there was something
wrong with the Jews." The writer Michael Crichton and the skeptic
Richard Lindzen freely compare climate science with Nazi race theory.

NASA climate scientist James Hansen went even further when he
described the trains loaded with coal heading for power plants as "death

trains—no less gruesome than if they were boxcars headed to crematoria, loaded with uncountable irreplaceable species." After a complaint from the Anti-Defamation League, Hansen later apologized. No such apology was forthcoming from Andrei Illarionov, an adviser to Russian president Vladimir Putin, who called the Kyoto Protocol "an interstate Auschwitz." Illarionov had previously rejected the phrase "interstate gulag" as too moderate, and, one suspects, still a little too close to the Russian bone. The best atrocities are those committed by other people.

Because climate change is a wicked problem, these metaphors then frame how we come to think about the issue as a whole. If we think of climate change as a ticking bomb, we see it quite differently than if we think of it as a fever, or a gamble, or a new Apollo space mission or a World War II battle. In each case, we imagine different causes, outcomes, and solutions.

But all of these framings are misleading. They encourage us to see climate change as a finite challenge that can be cured, overcome, or won rather than as an open-ended and irreversible condition that can only be managed. This shapeless multivalent issue readily takes on the form of the metaphors we apply to it and, as I will argue later, this can create a dangerous illusion of familiarity.

22

Communicator Trust

Why the Messenger Is More Important than the Message

IN 2007, GREG CRAVEN, A science teacher at an Oregon high school, posted a nine-minute video on YouTube with the teasing title "The Most Terrifying Video You Will Ever See." Craven is the unlikely successor to the seventeenth-century philosopher Blaise Pascal, who sought to "weigh the gain and the loss in wagering that God is." With his flipchart and marker pens, Craven weighs the pros and cons of action on climate change and concludes that we should all believe in it because, as Pascal said, if you gain, you gain all; if you lose, you lose nothing.

Craven, though, is no highbrow. He comes across as entirely relaxed, friendly, sincere, somewhat goofy, and utterly trustworthy, chatting in his light cheerful way with the kitchen clutter clearly visible behind him. Craven tells his viewers that in the Internet age, they can take immediate action by simply passing on his video to their friends. This simple admonition not only promoted his homemade video but also anointed it with the social proof of peer referral. It has now had more than six million views, making Craven one of the most successful climate change communicators of all time.

If words are frames and stories are the medium, then the person who communicates them becomes the most important and potentially the weakest link in the chain between scientific information and personal conviction. This sense of trustworthiness is a powerful bias and is entirely driven by the emotional brain and our intuitive ability to separate friends from foes. We ask ourselves: "Is this someone who I can trust on this issue?" "Is he or she honest and knowledgeable?" "Does this person appear to share my concerns and worldview?" and "Is this person approaching me with openness and friendliness?"

This is why, even though three-quarters of Americans still trust climate scientists as a source of information on global warming, they are nearly as inclined to trust television weather forecasters who are greatly less qualified as scientists but have a far more friendly, familiar, and approachable public profile. Unfortunately they also have a disconcertingly high level of climate denial. In a 2010 survey, only half of television weathercasters surveyed believed that climate change is occurring and more than a quarter believed that it is a "scam."

So what qualities are we looking for in a communicator we can trust? Integrity appears to be key, according to the political scientist Arthur Lupia. This, he argues, is directly related to our perception of what the communicator receives or risks from the communication. Thus we tend not to trust a communicator who is self-serving or unaccountable compared with someone who has taken a personal risk or would face penalties if he or she lied.

People who confound our expectations also become correspondingly more persuasive, especially when they cross sides, which we perceive as an act of great social risk and proof of their integrity. Whether regarded as freethinkers, whistleblowers, turncoats, or traitors, they are behaving contrary to expectations and invite our curiosity. What is more, they have a fascinating story to tell: of a challenging personal journey, often with a moment of revelation, a painful conversion, and the bravery of stating their new position despite the condemnation of their former friends.

Many notable deniers like to generate an environmentalist backstory. The Canadian denier Patrick Moore can, quite legitimately, claim to have been a founder of Greenpeace, even though he then spent the next twenty-five years undermining every campaign it ran. The high-profile Danish statistician Bjørn Lomborg claims that he is a "skeptical environmentalist" because he was once a member of Greenpeace (even though there is no record of his ever having been one).

And, going the other way, occasionally skeptics such as Richard Muller, the U.C. Berkeley theoretical physics professor, cross over to support climate change, gaining widespread publicity and credibility for their perceived independence. Muller now has direct access to senior CEOs, financiers, and policy makers simply by repeating what climate scientists have been saying all along.

The lack of trust in climate communicators is particularly acute among conservatives. The anti-warming campaigner Marc Morano identifies two critical mistakes: setting up the "highly distrusted, international bureaucracy" of the United Nations to supply the science (he chooses to forget that its real role is to only convene meetings of scientists) and choosing Al Gore, "the most polarizing figure in U.S. political history" as a figurehead. And that, he tells me with glee, is a total lose-lose! Morano is a highly combative character and clearly enjoys the thought that his side can claim the win-win.

He does have a point. Al Gore has provided a convenient hate figure for deniers. A word count of skeptical articles (among eight hundred op-eds) shows that nearly 40 percent of them mention Gore. He came with a long history of conservative mockery by George H. W. Bush as "ozone man" and is caricatured as self-serving, politically motivated, arrogant, and hypocritical—all qualities designed to damage the perception of trust. This can sometimes reach ridiculous lengths. The senior physicist William Happer argues that Gore is untrustworthy because the designer of one of his books has photoshopped clouds out of the cover photo of the earth.

However, the evidence suggests that distrust of Al Gore is a symptom rather than a cause of the conservative rejection of climate change. In fact conservatives' concern about climate change stayed fairly constant and even rose slightly when Gore's documentary *An Inconvenient Truth* was released and he received a Nobel Prize. If anything, their fall in concern came when his prominence declined. The issue appears to be less one of the presence of Gore than of the relative absence of other prominent climate communicators, especially those who speak to conservative values. In their absence, Gore has become an iconic figurehead for both the pro–climate change and anti–climate change campaigns.

Bob Inglis, former Republican representative for South Carolina, has been trying to fill the breach. His refusal to repudiate climate science in the 2011 election proved to be his undoing in the face of a concerted campaign by the Tea Party. However, the loss of his political career on a point of principle has given him an unparalleled legitimacy and trustworthiness in his new career as a campaigner for conservative action on climate change.

Inglis is passionate about the need for a conservative response to climate change, which he sees as a religious as well as political calling. Inglis is very Christian, very conservative (the American Conservative Union gave him a 93 percent approval rating), and strives to find evocative new language that will speak to conservative values, especially frames of respect, authority, and accountability. In his view, finding the right language is critical because, he tells me, "if you name the baby, you win it, and if you get the words around climate change right, you win the issue."

When people from the conservative political culture experiment with new language based on their values, the results can be surprising, intriguing, and, for many liberal environmentalists, appalling. They are going to have to recognize that opening up climate change to new communicators may also introduce new frames that make them feel decidedly uncomfortable.

New communicators like Colonel Mark "Puck" Mykleby, a lean, mean ex-Marine who translates sustainability into the language of special forces. He describes walkable communities as the future of American power and says that recovering from climate change impacts is like taking a gut punch and coming back swinging. Or Rob Sisson, the Catholic president of ConservAmerica, who argues that climate change is "the greatest single infringement on liberty after abortion" and that the pro-life campaign should fight for the six hundred thousand American babies who are "poisoned while still in their mothers' wombs every year by toxins released from burning fossil fuels."

But these are still high-profile professional campaigners—albeit with a new political stance. What climate change really needs are the voices of ordinary people who might not be fluent speakers or skilled orators but can bring an authenticity and genuine sense of common ownership to the issue.

They are rarely seen or heard—pushed to the sidelines by the experts or the advertising copywriters. It is shameful that high profile climate communications have never equaled the democracy of the "BP on the Street" commercials of 2004 that featured a range of opinions from ordinary people on energy and climate change, or the emotional power of the 2013 Chrysler Dodge Ram "So God Made a Farmer" commercial, composed entirely of photographs of farmers and their families.

Occasionally a small enlightened project seeks new voices. In the summer of 2012, Erik Fyfe and Albert Thrower toured the Northeast on a biofuel-powered motorbike filming interviews about climate change with people they met on the way: a barber, a distiller, a cranberry grower, and a sawmill operator among them. Some were worried, some were dismissive, and some were resigned, but the interviews, called "Slow Ride Stories," respected them all. Climate Wisconsin, a project of the state educational communications board, compiled short videos of timber workers, maple syrup farmers, fishermen, and a cargo ship pilot talking about the changes they have seen and their concerns for the future.

These interviews have a depth and human warmth to them that is so often lacking in the expert-driven discourse. The answer to the partisan deadlock and public disinterest starts, I am convinced, with finding new messengers rather than finding new messages, and then creating the means for them to be heard.

23

If They Don't Understand the Theory, Talk About It Over and Over and Over Again

Why Climate Science Does Not Move People

CLIMATE CHANGE IS A COMPLEX and technical issue that emerges from the theories, data, and predictions of scientific specialists. The problem is not just that scientists emphasize uncertainty and use obscure abstractions, but that they also often excise the very images, stories, and metaphors that might engage our emotional brain and galvanize us into action. Sometimes there seems to be a near-perfect mismatch between message and messenger—like those comedies where all the roles get mixed up: the homeless con man swaps places with the commodities broker, or the *I Love Lucy* episode where Lucy and Ethel get jobs in the candy factory while husbands Ricky and Fred stay at home.

This is not coincidental. The division between the emotional brain and the rational brain runs deep in our culture and is clearly expressed in the cultural divide between religion and science that first emerged during the European Enlightenment. It is, in the words of Tony Leiserowitz at the Yale Project on Climate Change Communication, "a long cultural mistake" to divide the two. They are, he says, inseparable, and "without that feeling of emotion, you cannot make good decisions. Scientists are human beings too, not Spock." Certainly so, and, as I will discuss later, the professional detachment of scientists leads us to seriously underestimate their own anxiety or despair.

This "long cultural mistake" continues to be expressed in the education system, which requires students to specialize far too early in the science, art, or humanities stream. This division then slices its way straight through the cultural debate around climate change. The vast majority of journalists, politicians, and environmentalists who dominate the communication of climate change have an arts or humanities background, and sometimes express their in-group contempt for scientists with a venom that would be considered entirely unacceptable in any other cultural realm—calling them "pointy heads," "white-coated prima donnas," or "mad cranks."

There are culture wars between scientists too, in particular the so-called paradigm wars between positivism (which uses experiments to establish findings that are declared to be universally true) and constructivism (which insists that knowledge is always situated in a time, in a place, and in a culture).

The few skeptics who have a legitimate scientific background invariably come from the positivist disciplines of physics, chemistry, and geology—particularly, it would seem, those with a background in the nuclear and petroleum industries. Their criticisms that climate science is being distorted for political or ideological reasons are reflections of deeper resentments about constructivism.

This all makes life very hard for climate scientists who find themselves torn between the need to maintain their professional detachment and engaging with this complex mental landscape of biases, narratives, and cultural prejudices.

The late Stephen Schneider, for many years America's best-known climate scientist, said that scientists found themselves in a "double ethical bind": "On the one hand, as scientists we are ethically bound to the scientific method. On the other hand, we are not just scientists but human beings as well. And like most people we'd like to see the world a better place. That, of course, means getting loads of media coverage. So we have to offer up scary scenarios, make simplified dramatic statements."

No scientist better exemplifies this struggle than climate scientist James Hansen, formerly of NASA, who told me that he does a "terrible job" at interviews and expresses a strong distaste for television. So much so that after his testimony to Congress in 1989 made international head-lines, he decided to "throw in the towel on the communication business" and hardly spoke in public for the next fifteen years until he felt morally compelled to. While we chatted in the autumn sunshine on the steps of Columbia University, we were constantly interrupted by admiring students coming up to shake his hand. He received their attention graciously but seemed shy and somewhat perplexed. There is no sense around Hansen—without doubt now the most famous climate scientist in the world—that he ever sought this profile.

As Bob Ward, communications director of the Grantham Institute for Climate Change at Imperial College London, says, "Our primary aim is to inform rather than to motivate. We perceive our role as providing factual inputs into rational decision-making processes." So when asked why people still do not get climate change, the vast majority of climate scien-tists will say that the problem is that people do not understand the research or the scientific process that produces it. For example, Jay Gulledge, a senior scientist at the Center for Climate and Energy Solutions, recommends explaining the theory, showing people the data, showing them the models, "and if it's hard to understand, you talk about it over and over and over again, because that's how you learn." This argu-ment is often called the information-deficit model, because it sees people, in the words of Ward, as "empty vessels who will respond appropriately once informed of the facts."

In recent years, with the continued lack of political or social action, scientists have become even more convinced that the problem lies with the distortion of this information by what they call "disinformation campaigns." This, they believe, is best countered with more information. Announcing a new report by the Potsdam Institute for Climate Impact Research, the research team leader declared his confidence that his data would "put to rest a misconception popular in some quarters, that global warming has slowed down." And of course, it did no such thing. Research shows that oppositional views can rarely be challenged effectively by new information and, if anything, are likely to be reinforced.

This was not the case when climate change first emerged. For the first ten or fifteen years, people's understanding of the underlying science was the single most powerful predictor of their willingness to change their behavior or support government policy. Successive Gallup polls during this period found that Republicans and Democrats who claimed to have a good understanding of the science were close together in their opinions on the issue.

But then the issue became polluted by political and cultural meaning. By 2010, when the number of people who said they didn't grasp climate change had fallen to just 3 percent, people's views were formed largely by their political orientation. Among Republicans, the more people knew about climate change, the less likely they were to believe in it. Overall, climate deniers had a slightly better general understanding of science than believers. As the Australian academic Clive Hamilton puts it, very elegantly, "Denial is due to a surplus of culture rather than a deficit of information."

Nor, I should add, is there any correlation between views on climate change and intelligence. In testing, law students with high IQs had no more interest in hearing different points of view than people with lower IQs; the only difference was that they used their intelligence to create more arguments for their existing view.

This confirmation bias extends to the trust we put into scientists as communicators. By and large, scientists are still well very trusted as a profession, even by Republicans and climate skeptics. However, when people are given a choice between scientists, they cannot resist extending that trust to the scientist who best confirms their own view.

Dan Kahan at Yale's Cultural Cognition Project ran an experiment in which he presented participants with different quotes on climate change ascribed to fictitious "experts." These were all described as professors at Ivy League universities and represented by clip-art photographs of middle-age white men in suits. Participants had no problems identifying which expert was the most authoritative and qualified to give an opinion. He was, of course, the one who presented the opinions that they already agreed with.

Ironically, one of the best proofs that information does not change people's attitudes is that science communicators continue to ignore the extensive research evidence that shows that information does not change people's attitudes. The vast majority of scientific communications is still in the form of data and graphs, and the main attempt to make it more appealing is to jazz it up with three-dimensional animated graphics and charts that whizz round, spin round, or bulge out.

The Intergovernmental Panel on Climate Change (IPCC) struggles with language. It does not have an easy job, considering the long, painful, and highly politicized process that approves its texts. As Sir John

Houghton, the founding chairperson, tells me, "Everything it says has to be agreed on, word by word, with one hundred people present, through simultaneous translation, and despite the efforts by Saudi Arabia and oil states to derail it." The language that survives this process becomes critically important for framing subsequent understanding of the science by politicians and the lay public. This is the key point at which the abstract rational language of probability needs to cross into the emotional language of threat.

Stephen Schneider understood this very well and urged the IPCC to adopt more colloquial language to convey degrees of certainty and uncertainty. For twenty-five years, the IPCC has been struggling with this task and, it must be said, has not managed it well, prompting strong criticism from a council of international science academies.

A team at the Department of Psychology at the University of Illinois tested the language used inside the IPCC reports and found that people severely underestimated the probabilities that the IPCC intended to communicate. The IPCC uses the term "very likely" to mean a chance of over 90 percent, but three-quarters of the lay readers put the odds far lower, some as low as 60 percent or less, even when they had the official IPCC definitions of these terms at hand.

Nonetheless, IPCC reports continue to use this language. When the fifth assessment report was launched in September 2013, the quotable press release banner headline was that scientists were "now 95% certain" that humans are the cause of climate change. A hundred years of retailing experience could have told them that people all too readily assume that a price tag of ninety-five dollars is significantly lower than one of a hundred dollars.

A similar problem emerges with the many attempts to quantify the scientific consensus that consistently finds that 95 to 97 percent of scientists or peer-reviewed papers agree human activities are responsible for climate change. If you are already distrustful, this statistic only reinforces the status of the 3 percent of dissenters. We have already seen that people are unmoved by numbers and that they focus on the individual with whom they can feel a personal connection. Narratives start to kick in that are deeply appealing to individualistic cultures that are suspicious of government and large institutions—of David and Goliath, the maverick, Joe the Plumber. A thousand people are a mob, but a single person standing up to that mob is a hero.

But while scientists are seeking to keep their language balanced and unemotional, they are missing the story that really would engage, excite, and inspire people: the story of themselves and their own passion for their science. After all, what truly engages the emotional brain are personal stories, and what convinces us of the trustworthiness of the communicator is our evaluation of his or her own commitment.

In his book *The Discovery of Global Warming*, physicist Spencer Weart says that he is going to tell "an epic story: the struggle of thousands of men and women over the course of a century for very high stakes. For some, the work required actual physical courage, a risk to life and limb in icy wastes or on the high seas. In the end they did win their goal, which was simply knowledge."

It's a bit frothy, I know, but it *is* exciting and makes you want to meet these scientists and hear what they have to say. I am fortunate to meet many climate scientists and I am always impressed by their passion, enthusiasm, humor, and commitment. Andrew Dessler, a professor of atmospheric physics at Rice University in Texas, argues that these personal stories are their strongest weapon and crucial for building trust. He regrets that "it goes against the culture of science, which emphasizes the collective and de-emphasizes the individual, but it's a skill climate communicators need to learn."

Indeed so. It is challenging, but then this is challenging for all of us. Despite the systematic smearing of scientists, they are still the most trusted communicators not just for their personal qualities but also for the quality of the scientific method they embody. There is no reason why they cannot present their findings and then take a step back and present their hopes, fears, and humanity. We do have a great deal in common after all.

24

Protect, Ban, Save, and Stop

How Climate Change Became Environmentalist

ON JULY 9, 1977, SHELDON Kinsel, a young lawyer working with the National Wildlife Federation, took the stand in front of a congressional hearing on energy policy. It was the first time that an environmental organization was entered into the congressional record on a topic that, at that time, still had no agreed name. Kinsel called it the "climate shift" and declared that "other *environmental* problems pale beside it." Even at this early stage, this was not just an *environmental* issue; it was the biggest *environmental* issue of all.

To begin with, environmental organizations simply expanded their existing activities to include climate change. A few years later, as the scientific and political momentum built, they clustered together in coalitions starting with the Climate Action Network in 1989 and, three years later, the Sustainable Energy Coalition. The movement built steadily and took off around 2005. There are now more than five hundred organizations campaigning on the issue.

Environmental organizations have always seen climate change as an obvious environmental issue. After all, it deals with atmospheric pollution, and this is exactly what they do. It emerged as an issue on the tail of a string of major legislative wins over air quality, lead in gasoline, ozone depletion, and acid rain. And climate change also contained a deeper critique of oil, coal, car-based transportation, high-consumption lifestyles, industrialization, economic growth, agribusiness, and meat eating—hot-button topics for many environmentalists.

Despite their differences—and it is a mistake not to recognize the vast differences in resources and politics between the rich Beltway groups and the grassroots environmental justice networks—all environmental organizations shaped climate change in their own image with narratives, images, and metaphors drawn from their previous struggles.

As they did so, another cognitive feedback ensued: The issue became more and more associated with the meanings that the environmentalists gave it, and any alternative framings became sidelined or remained unvoiced. Government, business, and the media were only too happy to help this happen—they all had their own motives for defining climate change as an environmental issue that they would take very seriously once they had dealt with more pressing matters.

Climate change became the biggest *environmental* issue of our times, reported by *environmental* correspondents in the media in special *environmental* reports, covered by *environmental* legislation created through *environmental* policy that was discussed in *environmental* speeches at *environmental* conferences.

This may create proximity for environmentalists, but for the wider public, these associations only make climate change more distant from their immediate concerns: as a luxury that can be kept on the edge of their pool of worry by economy, jobs, crime, and war. In wider polling, climate change always tracked concern about unrelated environmental issues, locked together firmly in the same cognitive frame.

And so an issue that requires an unprecedented level of cooperation has became exclusively associated with one movement and its various worldviews. Those who have historically distrusted environmentalism came to distrust climate change, and those who distrusted climate change came to distrust environmentalism all the more.

This is especially true of political conservatives. Former Republican representative Bob Inglis, struggling to develop a conservative narrative on climate change, is none too polite. "We conservatives tend to regard greens as gray-ponytail bed-wetters who've got their panties all in a wad," he tells me, laughing cheekily with the pleasure of saying this to a real-life panty-wadded bed-wetter. "And," he adds, "we think they are complainers, worrywarts. Listening to greens is like seeing a doctor who says, 'Oh my gosh, that's the *biggest* melanoma I have ever seen!'"

In a remarkable piece of circular logic, Myron Ebell of the free-market Competitive Enterprise Institute sees the environmentalist dominance as the ultimate proof that climate change is a fraud: "If it were an actual problem with real science behind it, then the environmentalists would not be able to take it over. It would have become a mainstream issue, and serious people would have discussed how to solve it."

Ted Nordhaus, the co-founder of the Breakthrough Institute, has built a reputation as an environmental heretic, though he prefers to call himself an "outlier" and "ecological modernist." He argues that environmentalists have proven to be "pretty well irrelevant," to have had "absolutely no impact," and to have "defined the space in such a way that it is very hard for other people to come in."

When I put these criticisms to Michael Brune, the executive director of the Sierra Club, he tries to force a laugh, but they clearly hit a nerve: "Isn't that ironic," he says. "They are saying that it is the environmentalists' fault that environmentalists are the only ones who are paying attention to this issue. I would say it's the fault of *every* segment of society that has not shown strong leadership on this. Everybody speaks for their own constituency, it's natural, and so of course we approach it using the words and images that motivate our own base."

I would not disagree, nor would I wish to dishonor the dedication of my colleagues who have worked so long and hard to maintain interest in this issue. But as my work has taken me away from my fellow greens into quite new groups—conservatives, Christians, Muslims, postal workers, coal miners, teachers, asylum seekers, delinquent teenagers, sheep farmers, Rotarians, custom-car enthusiasts, scouts—I have become aware of how poorly that environmental language works outside its own constituency. The problem is that in the absence of any competing narratives, these environmental words and images are so very—well, so very *environmental*.

Protecting, saving, banning and stopping things, for example. The Greenpeace offices in Washington, D.C., seem to be held together largely with posters and bumper stickers and almost all of them are about stopping things (Arctic drilling, greenwashing, nuclear power, logging, global warming, ocean destruction, CO_2, ExxonMobil, offshore drilling, Star Wars) or saving things (whales, Sumatran tigers, the biosphere, orangutans). Outside the toilets—themselves covered top to bottom with stickers—there is a banner (what else) from its Finance Department that reads, "Saving Time Saves Whales. Submit Your Receipts."

For linguistic expert George Lakoff, this language creates a false division. It creates the impression that the environment is some external entity that has to be protected or saved from an enemy that seeks to destroy it. Lakoff calls this a frame, but it is more recognizable as an archetypal narrative that has come to shape a large part of the environmentalist worldview in which people are divided into those who care and those who do not.

This leads, inevitably, to a judgmental streak. Al Gore famously said that people who believe the moon landing was staged and people who believe that the earth is flat "get together on Saturday night and party with the global-warming deniers." Considering that over a quarter of Republican voters are confident that climate change is *not* happening, it's going to be one hell of a party. This is a large constituency to offend.

Environmentalists are drawn to an anti-human rhetoric too, some of them talking about humans as a plague or virus that eats up the natural world. Chris Horner, Myron Ebell's fellow traveler in the Competitive Enterprise Institute, has a cute phrase: "Greens have the terrible toos: too many people, using too many resources." (To which one could reply that libertarian climate deniers have the "terrible frees": free markets and freedom from government.)

These generalizations do not reflect the diversity of the environmental movement or the continued work of many within that movement to build bridges with other constituencies. One major initiative, the Blue Green Alliance, was launched in 2006 by the Sierra Club and United Steelworkers and now involves fourteen of the largest U.S. unions and environmental organizations with a combined membership of fifteen million people.

The problem is that whenever environmentalists try to reach a wider audience, we cannot seem to resist reinforcing the frames that can lead other people to marginalize and ignore climate change. Consider this:

There is a low rumbling drone in the background (what documentary makers like to call "tone") while a string of very short-cut images flash across the screen: a hurricane, a turtle swimming over a coral reef, icebergs, a hurricane, thick smoke, burning oil wells, flies crawling over a dying child, rush hour in Mumbai. Now the drone has turned into a moaning choral lament and the images are flashing by: plucked chickens on a conveyor belt, a polar bear picking though garbage, rows of hogs in a meatpacking plant, a freeway interchange, skyscrapers, burning forests, an iceberg crumbling.

The title fades in: "THE IITH HOUR."

This film came out the year after Al Gore's *An Inconvenient Truth* and in many ways is a companion film. It is more of a montage—talking heads intercut with eye-catching images, and occasionally its creator, Leonardo DiCaprio, pops up in a storm drain.

It's a bit earnest but makes intelligent points and it works well for me. But then of course it would, because every single image, argument, and speaker screams ENVIRONMENTALIST, and I am one. The images are codes and, being one of the gang, I carry a mental copy of the codebook. Garbage = wasteful consumption. Polar bear = melting ice caps. Melting ice caps = global warming. Starving baby = a poor victim far away. Indian rush hour = overpopulation in a poor country far away. Skyscrapers = faceless corporations.

I also know that even the title *The 11th Hour* is a double metaphor. It says that humans are very new to the planet and that fossil fuel burning is very new to humans. And it's wheeling out the old plywood Atomic Clock again with the warning that we are approaching midnight.

These codes are another strong reason why so many people ignore climate change: The visual and metaphorical language that surrounds climate change marks it, irredeemably, as an environmental issue. These images, constantly reinforced in every news story and media item, create a tightly interlinked schema by which climate change is detached from the other issues (employment, economy, crime, defense) that people care most about.

And it is worse than that. Many of these images are not even about the environment but about the worldview of environmentalists. This is an important distinction. People with different values have their own codebooks, and they contain entirely different, and even contradictory, meanings. Lee Baringer, a former steel worker who now campaigns against climate change, recalls his initial hatred of the "small group of environmentalists in town who kept raising hell about the pollution from the steel mills. The air we breathed was truly foul, but to us it was the sweet smell of money because it paid the bills."

So, for many working people, meatpacking plants, factories, power plants, and traffic jams mean development and paid employment.

Multi-lane freeways mean mobility and freedom. When I lived in Taiwan in the 1980s, a fried chicken fast-food chain used to show videos of mechanized chicken-gutting operations inside its stores. For its Chinese patrons, these represented cleanliness and modernity.

Even as an environmentalist, I cannot watch those sweeping aerial shots of freeways, gleaming skyscrapers, and Las Vegas at night without feeling a surge of excitement about the glamour and bravado of the modern world. They may be dirty, but they *do* look fun. Just as I cannot watch that classic Christmas movie *It's a Wonderful Life* without feeling that the nightlife in Bedford Falls has greatly improved under the management of the revolting Henry F. Potter.

The people who campaign against action on climate change understand these different codings intuitively and build their communications around them. Back in 2006 the conservative D.C. think tank Competitive Enterprise Institute was frustrated by the publicity building up around Gore's *An Inconvenient Truth* and decided to produce a short video to put out its own view.

The thirty-thousand-dollar budget required a very simple in-house production. CEI general counsel Sam Kazman worked up a script and passed it around the office, and then they pasted it together with some generic stock footage and music. Kazman smiles when I ask him whether they tested it in focus groups. There was, he says, no testing of any kind. It was what they liked and, because it worked for them, they just put it out.

A further thirty thousand dollars booked just enough advertising space on cable TV that they could claim that this would be a national campaign. The real goal, of course, was to generate free media coverage in contrast to Gore's documentary. Kazman says that it "went viral before people even said things like 'went viral.'" It got the right-wing media but then went further: NPR, the BBC, National Geographic, even a quiz show. CEI got hate mail and angry phone calls. Greens hated it, but they still linked to it and passed it round.

The video starts with images of people sitting in a park on a summer's day, a jogger running down a beach, forests, and wild animals. Carbon dioxide, the voice-over tells us, is essential to life. We breathe it out. Plants breathe it in. It has freed us from a world of backbreaking labor, lighting up our lives, allowing us to move the people we love. As we hear this, we see images of Times Square at night and children being helped into the backseat of a car.

But there is a lurking menace. Over that ubiquitous documentary tone, the voice-over says that now some politicians want to label carbon dioxide a pollutant. Imagine if they succeed? What would our lives be like then? As these words are spoken, the images of Times Square and the children in the car fade into black.

Over the black the voice-over says, "Carbon dioxide. They call it pollution. We call it life," and the screen fades back to a young girl in bright sunshine blowing a dandelion clock into the wind. It leaves a lasting impression of the wonders of the life ahead for her. "I'd love to meet that girl one day," sighs Kazman.

It is devious, exasperating, and outright mendacious. But it is also damned good communication. So *maddeningly* good that I too have become an unwitting vector in its transmission—inviting people in my communications trainings to watch it and learn how to construct a narrative around positive universal values. It is a textbook example of how to speak directly to the emotional brain. As Fred Smith, the CEI founder, says, "It should always bring a tear to your eye."

The video is an artful compilation of frames for life, civilization, health, safety, hope, and salvation. And by contrast, the image of Times Square and the children fading into darkness speaks equally well to metaphors for decay and death—as it would in every culture in the world.

Which makes it curious, to say the least, that the World Wildlife Fund uses the same metaphors at the core of its largest public engagement exercise around climate change, Earth Hour. Every year it encourages us to turn off our lights for just one hour. Its website shows clips of the lights turning off in entire cities and national monuments: the Eiffel Tower, the statue of Christ the Redeemer in Rio, and, yes, New York's Times Square too.

Changi Airport in Singapore is also a regular participant in Earth Hour. It dims its lights for an hour and boasts that this saves as much electricity as would be used in an apartment over three months. A fine achievement if one forgets the emissions of the 70,000 planes that land there during the same period.

Earth Hour co-founder and executive director Andy Ridley stresses that it's not about saving power: "What it is meant to be about is showing what can happen when people come together."

So this is a huge symbol. Politicians like it because they love big, cheap empty gestures. U.N. Secretary-General Ban Ki-moon offers his support with the words "Let us use 60 minutes of darkness to help the world see the light." British prime minister David Cameron tells us that Earth Hour is "a huge symbol of global solidarity, an inspiring display of international commitment."

The World Wildlife Fund thinks it is a huge success—a small, simple, visible act that generates a social norm, and is, let's face it, a wonderful vehicle for its public profile and fund-raising. And of course, many thousands of committed environmentalists express their solidarity in spontaneous Earth Hour events, sitting in the dark and feeling their sense of shared involvement.

But there is no avoiding the fact that, if one is going to play in the world of symbols, one had better get it right. However you read it, a universal frame for decline, decay, and death is being promoted on a vast scale all around the world as a symbol for climate change.

Deniers understood this immediately. The blogger Alan Caruba posted an aerial photograph of an entirely dark North Korea at night with the caption "It's Always Earth Hour in North Korea. Electricity is the difference between the Dark Age and the present age. People who hate civilization are welcome to live out in the wilderness and burn dung to cook their meals."

The real problem, though, is that Earth Hour is not in the least interested in what Caruba or his sympathizers think. It makes sense to people who share the values and can read the codes (solidarity, shared commitment), and that is good enough. Environmental messaging around climate change is not deliberately exclusive. It would *like* to reach other people, but because it is not interested in reflecting other people's values, it keeps takings bricks out of the bridge that unites people around a common interest in their future.

25

Polarization

Why Polar Bears Make It Harder to Accept Climate Change

IT'S A GRAY AUTUMN SUNDAY afternoon and I am following Aurora, a giant animatronic polar bear, the size of a double-decker bus, as she stalks the streets of London, pacing, sniffing, roaring. Around her neck are strips of cloth bearing the names of three million supporters. She is preceded by three ice spirits with glitter and wands doing some kind of free-form dance. All around me are people in white face paint with blackened noses, woolly white hats, and stick-on ears.

Now, I know that this is a Greenpeace protest against drilling in the Arctic, but without my codebook, a forty-foot-long icon bearing three million blessings being pushed by two thousand people looks more like the annual procession of the Hindu chariot of Lord Jagannath—known to the world as the Juggernaut. I wonder if we should throw ourselves under its paws.

Of course this is not a religion and it would be silly to say it was. But nor it is a carnival or a big press stunt either. It is revealing that when I invite the people in the march to tell me about the puppet, they don't talk about *it* at all—they immediately start talking about their concerns about polar bears, the climate, and the future. To them Aurora is a powerful representational symbol onto which they can project their collective values and concerns.

Polar bears are the ubiquitous symbol of climate change. The maga-zine cover, B-roll footage, stick-it-in stock photo agency picture for any climate change story. When *Time* magazine ran its first special issue on climate change, its cover headline, "Be Worried. Be Very Worried," accompanied a photograph of a lonely polar bear perched on a tiny iceberg looking at us with anxiety, hunger, desperation . . . well, with whatever expression we wish to project onto it.

As a result, people in focus groups choose the polar bear as the number-one icon of climate change, saying that they are drawn to it because it represents "the idea of a pure fragile environment most affected by change."

No environmentalist I spoke to could ever recall a formal decision in their organization to select a polar bear as a campaign icon. The National Wildlife Federation justifies the emphasis on polar bears because they are "the proverbial canary in the coal mine" but this is a weak (and zoologically confused) metaphor. The long-term future is not looking good for polar bears but there are very large variations in predictions for them in the short term, with some populations (especially around Hudson Bay in Canada) declining and others rising following the suspension of hunting.

The real reasons that polar bears became an icon are that climate change initially focused on the Arctic and that environmental organiza-tions have always used iconic megafauna to symbolize complex resource issues. No other progressive campaign constituency would have chosen this emblem—there are no bears of any kind in the materials of human rights, refugee, health charities, trade unions, business organizations, or faith groups. The British development charity Christian Aid even attacked the icon in a poster that showed Africans on cracked earth on one side and cracking ice on the other, with the caption "Climate change threatens more than just polar bears and ice caps."

More basic organizational dynamics are at work too. Like dolphins and panda bears, polar bears combine excellently with the dynamics of large-scale fundraising. Different organizations compete for the best package of spin-off merchandise, tote bags, a "Save Our Ice" water bottle, a "Celebrate Mother's Day" photo of mom and her two cubs. A large dona-tion to the National Wildlife Federation earns you a four-foot-high stuffed polar bear of your very own.

And they make great costumes, no small consideration for protests that are less about confrontation than about generating a media image.

So polar bears march with a placard reading "You fly, I die" or sit holding a begging cups with a sign reading "homeless."

The biggest problem with polar bears, though, is that they play so poorly to our cognitive biases. An issue that suffers from a lack of proximity has chosen as an icon an animal that could not be more distant from people's real life. Indeed, outside a zoo, people are far more likely to have seen an activist in a polar bear costume than an actual polar bear.

Maybe ice works better. Certainly, it does tap into the mental model that we all have from our own experience, that ice melts on a sunny day. But you cannot show an absence without showing a presence, and a large block of melting ice is still a large block of cold ice. There is a reason why ice, penguins, and polar bears appear regularly in advertisements for freezers rather than ovens. After all, if we were running a campaign against global cooling, would we think it reasonable to have the camel as its central image?

Semiotics is the name given to the interpretation and study of nonlinguistic signs, such as images and icons. Judith Williamson, a pioneer of the study of the semiotics of advertising, says that there has been a recent "avalanche" of artworks—novels, poems, paintings, photographs, and advertisements—that celebrate snow and ice, polar bears, penguins, and glaciers.

She argues that this focus on what is vanishing means that we are perpetually looking backward rather than forward, gazing at what might be gone rather than at what might come into being. It is a visual iconography that speaks of loss, and is tinged with melancholy.

"How powerful an image ice is for slow, dripping loss," Williamson says. "How many of our emotions are frozen too, along with this imagery! We stand at the brink of something, hoping it can be prevented. The idea of preserving the glaciers and polar bears channels the wish to freeze the world as it is, to hold on, not to let things go."

26

Turn Off Your Lights or the Puppy Gets It

How Doomsday Becomes Dullsville

IN THAT CREEPY WAY WE know from horror films, the camera glides along a darkened landing toward an opened door. Inside we can see Dad, cuddled up in bed with his daughter, holding open a storybook.

"There was once a land," starts the father, "where the weather was very strange. There were awful heat waves in some parts and in others terrible storms and floods. Scientists said it was caused by too much CO2, which went into the sky when the grown-ups used energy, and the children of the land would have to live with the horrible consequences."

As he speaks, we see the pictures moving around in the book. A black cloud of carbon dioxide with an angry face forms in the sky. Lightning strikes. A cute cartoon puppy waves one last time before going under the rising waters. Hang on, we start thinking, this isn't any ordinary bedtime story (and clearly, many people reckoned later, this isn't any ordinary dad to be reading a kid this stuff).

The little girl looks up with her big, moist eyes. "Daddy," she asks, "is there a happy ending?" Cut to black. "It's up to us how the story ends," the voice-over says and tells us to go to a government website called Act on CO2. But the actual message is pretty clear: Turn off your lights *or the puppy gets it.*

This nine-million-dollar advertising turkey was launched on October

9, 2009, during the commercial break halfway through Britain's most popular soap opera. Within a week, it had been pulled, after the Advertising Standards Authority, an independent agency that polices advertising, received more than nine hundred complaints about the commercial being political, misleading, and frightening to children.

Communications specialists marveled at the creativity with which this campaign broke every recommendation they had ever made to the government on how to talk about climate change. Clearly the brief had included a requirement to be as depressing, judgmental, manipulative, untrustworthy, and condescending as possible, but the absolute genius—for which the advertising agency surely deserved some award—was framing a deeply contested scientific issue as a children's fairy tale. "It's utter rubbish," said Ed Gillespie of Futerra Communications, which had previously advised the government to communicate a positive and aspirational vision. "It is about as much use as a marzipan dildo."

The reaction to this ill-conceived campaign goes to the heart of a debate that has reverberated through every report, documentary, and article since the very first warnings of climate change: To what extent should communications concentrate on climate change as a disaster?

Cognitive specialists suggest that we need to feel that climate change is a *dread risk* before we will act. Professor Elke Weber of Columbia University's Earth Institute writes, "It is only the potentially catastrophic nature of rapid climate change and the global dimension of adverse effects that have the potential for raising a visceral reaction to the risk."

In conversation, many experts are of the view that it would be dishonest not to tell things how they are. The pioneering investigative journalist Ross Gelbspan says that his professional business involves telling the truth and, sadly, the truth about climate change really *is* apocalyptic. The Australian academic Clive Hamilton tells me that environmentalists constantly confront him with homilies about his pessimism. When they demand that he be more hopeful, they are, he feels, coping with their own kind of denial.

The problem, as the reaction to the advertisement showed all too well, is that when people feel threatened and isolated, they can adopt a range of strategies to diminish their sense of internal fear: denial, uncertainty, playing down the threat, fatalism, and anger toward the communicator. Psychologists call these responses maladaptations, in that they are responses that do nothing to reduce the actual level of risk. The wider research into fear responses shows that people can also become desensitized and may require ever greater urgency or threat to stay interested.

For this reason, some communications specialists argue for a balanced narrative that starts with a positive vision to offset the bad news. There is a cognitive basis for this approach. Because the emotional brain leads in decision making, its initial impressions will sway subsequent decisions. In experiments, people are far more inclined to agree to make personal sacrifices if they have first been invited to generate, in their own words, the longer-term positive reasons for changing their consumption.

Dan Kahan of Yale's Cultural Cognition Project, though, is none too impressed by this approach, which he calls the "goldilocks dialectic," because it suggests that people can be motivated by "neither too much alarm nor too little but just the right mix of fear and hope." He stresses that the perception of risk is formed by the norms within social groups and that effective communications need to respond to these values, rather than seeking some perfect cocktail.

The communicators, too, are subject to their own cultural biases in the way that they construct disaster narratives. Although they are simply trying to motivate people by stimulating a sense of "dread risk," they cannot avoid shaping the message around their own values and worldview. The psychotherapist Sally Weintrobe wonders whether apocalyptic messaging is a coping mechanism for the communicators who are unwittingly projecting their own anxieties onto the people they want to engage.

And the deniers *are* right about some things: There really *is* a lot of disaster stuff around environmentalists.

The rise of the environmental movement is all bound up with the lexicon of apocalypse. The words *doomsday* and *apocalypse*—despite their ancient religious origins—really take off in popular usage only in the 1960s. As one skeptic (a physicist, of course) writes, for "warmers" it is Halloween all year long.

Here, for example, is the environmental novelist John Atcheson with his own "I have a dream" speech: "Imagine a world where vast regions of an acidic ocean are dominated by jellyfish. The land? An unending series of drought, flood, fire and famine. The coasts will be their own special blend of hell on earth." Sounds super. And there is much, much more where that came from.

People who hold the world to be just, orderly, and stable have a deep-seated loathing of this kind of apocalyptic messaging. The idea that they could be subject to arbitrary impacts upsets their belief that the worthy are rewarded and that only wrongdoing is punished. In an experiment when people with these "just worldviews" were presented with apocalyptic messages, their belief in climate change fell dramatically.

One of the videos they were shown in the experiment is worth mentioning in detail: A man is standing on a railroad track as a speeding train comes ever closer behind him. "Some people say the irreversible consequences of global warming are thirty years away," says the man. "Thirty years? That won't affect me." He steps off the track, and we can see that hidden behind him all the time was a young girl. With the train now mere feet away, up comes the slogan "There's still time—fight global warming."

This television commercial was made in 2006 by the Environmental Defense Fund. I doubt that anyone other than dedicated greens (and possibly even not them) would respond well to it. Certainly it is powerful and grabs one's attention, but it generates no sense of efficacy—the best thing you can do in the face of a runaway train is to jump the track, pulling the little girl behind you.

I have already shown that people evaluate new information in the light of their recent experience. In rich Western countries they will have little available experience of environmental or social collapse, but they will have a large mental library, often in the form of polished stories, of failed *prophecies* of collapse. It is these that most readily come to mind.

The greatest, and longest-sustained, postwar fear was of imminent nuclear apocalypse, which then morphed seamlessly into the fears of nuclear power and the early environment movement. As an article in the *Age*, a skeptical Australia newspaper said, under the heading "A Climate of Fear," "The Bomb was back, like the ghost at a banquet of anxiety. From Al Gore to the Intergovernmental Panel on Climate Change, everyone had grim news for the planet."

Environmentalism has its own unfulfilled prophecies. The Club of Rome's *Limits to Growth* report, which became a founding text of modern environmentalism, predicted imminent global collapse from "overshoot." It included, even in 1972, two pages dedicated to climate change. It sold an astonishing twelve million copies and prompted a torrent of pessimistic prose. *Time* magazine's story on the report, headlined "The Worst Is Yet to Be," told us that "in Los Angeles a few gaunt survivors of a plague desperately till freeway center strips, backyards and outlying fields, hoping to raise a subsistence crop."

And still the warnings come. There is mad cow disease, bird flu, swine flu, Y2K, Iraqi weapons of mass destruction, terrorist scares that never eventuate, and various health, nutritional, and economic "time bombs." All of these are entirely plausible and many are as valid as ever. It is just that they never seem to happen when anyone says they will, and people have only a limited capacity for staying on red alert.

The commonly used story for failed prophecies, which has appeared regularly in the climate denier discourse since the early 1990s, is the fable of Chicken Little, who, after being hit on the head by a falling acorn, persuades her fellow animals that the sky is falling. The local wily wolf exploits these fears and persuades the gullible animals to find safety in his cave, where he then eats them.

This is actually an ancient story that first appeared in Buddhist scriptures some 2,500 years ago. It has been constantly reinvented with updated moral conclusions to fit the values of each society that tells it: the evils of mass panic (India), the need to find your own evidence (Tibet), individual stupidity (Europe), and in a 1943 Disney cartoon, the dangers of wartime rumors. In all of its manifestations, though, it is a marvelous display of the social norm at work, and how, though repetition between peers, a lie can gain social proof. It is, then, a story that is highly relevant to climate change.

But the more appropriate metaphor for failed climate change predictions may be Aesop's fable of the boy who cried wolf. A boy who is asked to guard over the sheep repeatedly cries that the wolves are coming. After he has been found to be lying three times in a row, no one pays any further attention. Then the wolves really do come. Aesop says that the moral is, don't lie, because no one will believe you when you are telling the truth.

Conviction in climate change depends, like the warning of the wolf attack, on the perceived authority and honesty of the communicator. No warning will be believed once the trust in the communicator is lost. But timing is also essential—after all, the boy was not exactly lying; the wolves really *were* there, they *were* a real risk, and they *did* come and eat the sheep, but just not when he said they would.

In 1989 NASA climate scientist James Hansen asked, "Must we wait until the prey, in this case the world's environment, is mangled by the wolf's grip?" He disagreed with the wider scientific community that a few cool years might discredit the whole issue. "The time to cry wolf is here," he said.

Hansen has been proven both right and wrong. He was right because "crying wolf" put the issue high on the international agenda. But the concern that a few cool years could undermine trust was well founded. Surface temperatures appear (although the measurements are debated) to have leveled off since 2000. Conveniently ignoring the complex scientific explanations of what might be happening, this has been widely reported to suggest that the original fears were without foundation, often by the same newspapers and news programs that reported the original fears under lurid exaggerated headlines.

The news media is fickle and inconsistent in its coverage. It happily stokes up fears on the tried and tested principle that "if it bleeds, it leads." A study of news stories about a 2006 Intergovernmental Panel on Climate Change report found frequent use of the adjectives *catastrophic, shocking, terrifying,* and *devastating,* even though not one of these words was present in the original document. It is not surprising that 40 percent of Americans believe that the media exaggerates the threat of climate change.

The climate scientist Michael Mann sees this overstatement as part of the media cycle that then feeds denial. "Saying climate change is really bad became stale. And so journalists felt they had to find a new narrative that they ironically had helped create, that the science had somehow been overstated."

Talking to scientists, though, I sense that they are not just concerned with the distortion of the data but also feel deeply uncomfortable with the construction of the disaster narrative itself. Their definition of professional integrity requires the production of a well-balanced product stripped of all emotional hyperbole, and seeing it reconstituted this way is as insulting to them as dousing a dish with hot sauce would be to a master chef.

Regardless of their professional pride, there is no easy answer to the question of how to best communicate the serious threats contained in their science in a way that people respect, understand, and heed. Solutions are likely, as usual, to lie in a plurality of approaches, with different communicators speaking in different ways to different audiences. As Dan Kahan argues, people interpret these warnings through their lens of their values and cultures. Some people want the ten course scientific tasting menu and some just want a taco with lots of hot sauce.

The problem is that, thanks to the pervasive climate silence, they are rarely offered either option. Those of us who work in the field assume that everyone is talking about climate change and that, as stated in a paper by three Australian psychologists, "the nonstop media coverage of the climate change threat, its often apocalyptic portrayal and ubiquitous images have all given the threat of climate change a very substantial virtual and psychological reality."

That might be true if there really was nonstop media coverage. But, as the blogger Joe Romm points out, in the United States, there is virtually no presence of any kind for climate change in mass popular culture, with the very rare coverage on mainstream news sandwiched between millions of dollars of fossil fuel ads and trivia. The conclusion, again, is that what is not said is as important as what is said.

In May 2011, for example, the three main networks gave blanket coverage to the marriage of the future king of a minor island (my own as it happens) rather than the record collapse of ice cover in the Arctic. Climate change still hardly figures against the reporting of the economy, terrorism, or any other pressing concern. And in reality, most people aren't watching that either: They are switching over to the game shows, soap operas, or shopping channels, or watching funny animals, not drowning ones.

27

Bright-siding

The Dangers of Positive Dreams

THE CLIMATE APOCALYPSE NARRATIVE HAS a positive, optimistic, and altogether bouncier twin: the bright side. It provides a far more coherent narrative than apocalypse and so can be summed up in a paragraph:

> Climate change is a challenge, but it is also a great opportunity. Humans have always triumphed over adversity and come through stronger. Our ingenuity, technology, and capitalism have created unbelievable progress and will continue to do so. We can be anything we want to be, so the real enemy is negativity and despair, which must not be allowed to poison this positive vision.

In cognitive terms, the bright side appeals directly to the emotional brain and passes through its biases with flying colors. It replaces uncertainty with confidence. It replaces short-term sacrifice with the offer of immediate rewards of wealth and status. And it compensates for the taint of failure and self-doubt that hangs around climate change with an overstated confidence in technology and economic growth.

Americans are particularly prone to bright-siding. The journalist Barbara Ehrenreich, who helped popularize the word, argues that "positivity is a quintessentially American activity, associated in our minds with both individual and national success, but it is driven by a terrible insecurity which requires a constant effort to repress or block out unpleasant possibilities and 'negative' thoughts."

The bright side is the narrative of choice for business entrepreneurs and politicians and reflects their disposition to risk taking and optimism bias. As Barack Obama said during a 2008 presidential debate, "This is not just a challenge; it's an opportunity."

This balance between challenge and opportunity is nicely represented for bright-siders by the Chinese ideogram for "crisis"—*weiji*, which, they like to claim, combines the character for "disaster" (*wei*) with the character for "opportunity" (*ji*). And here they are neatly tethered together: Every crisis, they argue, offers both a disaster *and* an opportunity.

Following the inspiration of John F. Kennedy, who used it in his 1960 presidential speeches, Al Gore used *weiji* as the central theme in his testimony to the U.S. Senate on climate change and then in his book *An Inconvenient Truth*. The characters for *weiji* dominate the front cover of *The Death of Environmentalism*, a revisionist critique of environmentalism by the Breakthrough Institute. The report's central theme, not surprisingly, is the ineffectiveness of environmental doom-mongering.

However, the inconvenient linguistic truth is that *ji* doesn't mean "opportunity" at all—it simply means "a moment," and in other contexts it could mean "airplane" or "inorganic chemistry," both of which make an even better metaphor for climate change.

Bright-siders are particularly prone to this kind of selective quotation. Ali Al-Naimi, the Saudi Arabian minister of petroleum, responds to concerns about climate change with a quote from bright-side icon Winston Churchill: "A pessimist sees the difficulty in every opportunity; an optimist sees the opportunity in every difficulty." Martin Luther King Jr. is another model of charismatic leadership, and they frequently point out that his appeal to the American conscience started with the upbeat words "I have a dream" not "I have a nightmare."

Bright-siding, then, is not just an alternative view; it is a narrative antidote to the negativity of the apocalypse, in which the real problem is pessimism itself. Loving quotes and strong leadership, bright-siders frequently cite FDR's inaugural speech in 1933: "The only thing we have to fear is fear itself."

The senior environmentalist David Orr is having none of it and parries the quotations. "Happy talk," he says, "was not the approach taken by Franklin Roosevelt when facing the grim realities after Pearl Harbor. Nor was it Winston Churchill's message to the British people at the height of the London Blitz." Instead, he says, leaders told the truth honestly, with conviction and eloquence.

Climate bright-siders have a homeland too—Sustainia, a Copenhagen-based initiative to visualize the sustainable future based on "a new narrative of optimism and hope" that "seeks to inspire and motivate instead of scare people with gloom and doomsday scenarios." Not surprisingly, "I have a dream" figures prominently in its materials.

Sustainia even has its own language, Sustainian, in which it describes its "land of exciting possibilities, smart people, and positive images." Sustainia is more than a concept; "it is a luxurious, desirable lifestyle you cannot live without."

A promotional animated video gives us a window into the irresistible luxury of Sustainia's virtual world. The solar panels and revolving windmills are glinting in a late-summer afternoon, and the light rail tram glides silently past the ivy-clad condos, as the beautiful people, with rippling muscles and strangely distended limbs (with no whining children in sight), dine alfresco on their balconies or frolic in the long grass.

I spoke with Laura Storm, the wonderfully named executive director of Sustainia, a week before she was due to have her first baby, which, in fluent Sustainian, she described as "an exciting new chapter in my life." Sustainia developed, she said, to provide businesses with their own narrative around climate change that could inspire them and their employees. She described Sustainia as "a translation machine," noting that "environmental organizations have played a huge role in building awareness of climate change, and Sustainia revitalizes this with a narrative of what is possible"—which, increasingly, means working directly with industry to profile practical solutions.

In this spirit, every year Sustainia recognizes these solutions in a high-profile awards ceremony presided over by the ultimate bright-sider, Arnold Schwarzenegger, one of the few people on earth who actually looks like the bizarre avatars in Sustainia's virtual world. After handing out the Sustainia awards to the "new breed of green solutions action heroes" in 2012, he said, "Being a champion in bodybuilding, in movies, or in politics, I learned that the key to success is motivating and inspiring everyone to be a part of the solution, not just part of the problem."

Sustainia's glossy materials tell the story of a future action hero, Prabhu, a Seattle-based solar panel venture capitalist flying to his business meeting on his solar plane with "lightweight solar panels affixed to each wing." In Sustainia, we hear "there is no need to feel guilty about the amount of energy you use—you know it is nonpolluting. Ahhhh . . . fresh air." *Fresh*, with its companions *clean* and *bright*, is a key word in Sustainian and the bright-side vocabulary.

One of the leading partners in Sustainia is the Climate Group, an international network of business and government leaders promoting a "low carbon future that is smarter, better and more prosperous." In particular, it talks constantly of the clean revolution—not, you understand, that old fashioned *dirty* revolution of class struggle, social justice, and sweaty unshaven thugs in bandannas; this is the *clean* revolution of open-shirted executives working on their laptops in transit lounges.

Zachary Karabell is one of them: a financial manager and pundit who has built his career from being, in his words, "optimistic about everything." No doubt this is why the World Economic Forum calls him a "Global Leader for Tomorrow." Karabell argues that we have "chosen the path of pessimism instead of a path of innovation." Climate change, he says, is not the disaster we fear but instead "one more obstacle that humans can meet, one that may spur innovation and creativity." This will, he says, "make us tough."

This inspiring vision conveniently ignores the billions of people who find life quite tough enough already and for whom "one more obstacle" could destroy their lives.

However, the narrative of technological solutions and elitist control that permeates bright-siding can all too readily embrace a darker vision of the future in which vast planetary-scale engineering solutions (usually known as geo-engineering, or more chillingly, *climate remediation*) remove carbon from the atmosphere or reflect sunlight away from the earth.

These technologies are fascinating to entrepreneurial billionaires such as Niklas Zennström, cofounder of Skype, and Bill Gates, who is described by *Fortune* as "the world's leading funder of research into geoengineering."

In 2007 Al Gore and Sir Richard Branson, billionaire founder of Virgin airlines, announced a competition called the Virgin Earth Challenge offering twenty-five million dollars "to find commercially viable designs to permanently remove greenhouse gases from the atmosphere." Curiously, the eleven finalists were announced in Calgary, Alberta, under the patronage of the Canadian tar sand industry.

While turning up the optimism about technological solutions, bright-siding turns down the volume of threat. It is only a few more notches on the dial before one is deep into outright denial. The libertarian Cato and Heartland Institutes, while strongly denying the existence of climate change, have no problem promoting the solution of geo-engineering as "more cost-effective" and "delivering measurable results in a matter of weeks rather than the decades or centuries required for greenhouse gas reductions to take full effect."

The fact that outspoken free market deniers can endorse the positive solution, while refusing to accept the existence of the negative problem, is revealing and relevant. Bright-siding is ultimately a regressive narrative that validates existing hierarchies. It promotes an aspirational high-consumption lifestyle while ignoring the deep inequalities, pollution, and waste that make that lifestyle possible. And this is why, despite its upbeat tone, it is just as unappealing to many ordinary people as the apocalypse it seeks to replace.

28

Winning the Argument

How a Scientific Discourse Turned into a Debating Slam

I'M SITTING IN THE CAFE in Union Station in Washington, D.C., watching Marc Morano demolish a plate of oysters. While Daniel Kahneman managed to make drinking tomato soup into a Taoist meditation exercise, Morano has found a way to ingest oysters without stopping talking, like a grazing shark. And boy, this fellow can talk. He is, in his words, a motor-mouth: loud, highly opinionated—I find him kind of fun and kind of appalling in equal measure.

Morano is one of the key communicators in anti-warming activism, constantly in demand for TV and other media and also as an informal media service to enable access to other deniers. So I was interested in obtaining his perspective as a communications specialist, rather than a campaigner, on how he shapes the story to make it more appealing.

His strategy is entirely reactive, waiting to see what evidence and narrative the other side comes up with and then, he says, "knocking it down." The journalist Ross Gelbspan, who has interviewed almost all the professional contrarians, has concluded that they are "detached internally from the substance of what they are naysaying and motivated by the gamesmanship of showing how clever they are—as though it is all a game of chess." Though, in Morano's case, it is a more aggressive contact sport that comes to mind.

Not surprisingly, Morano thrives in the sound bite debate format beloved of live television and is promoted to the media by his employers, the Committee for a Constructive Tomorrow (CFACT), as a "credentialed counter guest" who will "offer a lively, fair and balanced discussion."

There is no question that he makes great debate television using his remarkable recall to gun down opponents with a blast of citations. Andrew Watson from the University of East Anglia Climate Research Unit was so wound up after a live television debate with Morano that he forgot that his microphone was still on and could be very clearly heard saying, "Christ, what an arsehole." Morano loves this: "Sure," he says, "Morano's an A-HOLE!"

Morano's primary strategy is to play the enemy narrative and to discredit individual climate scientists. He says he likes to have fun, to go after somebody who is being particularly silly and ridiculous, and to use mockery. But it didn't sound so fun when he wrote, on his blogsite, "We should kick scientists while they're down. They deserve to be publicly flogged." He laughs it off. "Come on," he says, "it was a stupid *expression!*"

Scientists, especially those who have received a torrent of abusive and threatening e-mails after being featured on Morano's blog, say he incites bullying and intimidation. When I tell one scientist that I have met Morano, he says, "I hope you had a shower afterward." Another calls him "a vicious, nasty man" and says he would be "a leading Nazi propagandist—a storm trooper or worse." Both are very concerned that I should not use their names.

Michael Mann, the much-abused director of the Earth System Science Center at Pennsylvania State University, is now so battle-hardened that he doesn't care anymore whether I use his name or not. He calls Morano a "hired assassin" who "spreads malicious lies about scientists, paints us as enemies of the people, then uses language that makes it sound like we should be subject to death threats, harmed, or killed."

Morano is unapologetic. All he does, he says, is to post the contact details that he finds on the public websites of scientists and to suggest that his followers tell them what they think. In his mind he is performing a public service. "Scientists live in a bubble," he says, "and this is the first time that they ever hear from the public. It's refreshing. It's healthy. It's good for the public debate."

Scientists welcome debates with their peers, especially the respectful exchange of views between experts at conferences and meticulously referenced responses in peer-reviewed journals. What they hate are the pure performance battles that can be won by the communicator who puts up the best fight. This is why, more than anything, televised debates are what campaigners like Morano want and need. These transform the complex "wicked" issue of climate change into a simple narrative of competing sides. As he knows full well, the mere existence of a debate is enough to persuade people that climate change is still debatable.

But debates are not just a fight. They reflect, in miniature, the process by which people come to form their views on climate change—enabling them to weigh the trustworthiness of the communicator, the social cues, the "narrative fidelity" and, finally, to pick a side.

On March 23, 2007, National Public Radio and fifty affiliated radio stations across the United States broadcast a debate between high-profile scientists and skeptics on the topic "global warming is not a crisis." This debate was unusual because the audience held a ballot before and after the debate, making it into a large, albeit crude, experiment in testing different presentation styles and narratives.

The vote showed that before the debate, the advocates of mainstream science were comfortably in the lead. By the end of the debate, they had lost a third of their support. They had the best-qualified team and a position supported by every scientific institution and government in the world. And somehow they lost.

The decisive factor that won the debate for the skeptics appears to be their use of stories and social cues reinforced by humor. The skeptic Philip Stott, a retired professor of geography from London University, did not just say that climate is complex; he said it was as chaotic as Glasgow on a Saturday night, and that understanding it is like "trying to play Mozart's wonderful Sinfonia Concertante 364 when you've no viola part and only a quarter of the violin part."

While Stott played the English gentleman, his teammate, the novelist Michael Crichton, demonized the messenger. He posed a question: "Haven't we actually raised temperatures so much that we, as stewards of the planet, *have* to act?" Then, without missing a beat, he added, "These are the questions my friends like to ask as they get on board their private jets to fly to their second and third homes." The audience roared with laughter. Crichton returned regularly to these hypocritical environmentalists who, he said, buy a Prius, drive it around for a while, and give it to the maid.

The advocates tried hard with metaphor. Gavin Schmidt of NASA talked about scientists being like crime scene investigators—skilled experts tracking down the killer. Richard C.J. Somerville, a professor of meteorology at the University of California, San Diego, argued that choosing not to fight global warming is about as irresponsible as not making payments on a high-interest credit card.

The social cues were already clear. Advocates are judgmental elitists and hypocrites. Skeptics are relaxed and can enjoy life and have a laugh. As the famous election question has it: Who would you rather have a beer with? The man who wants to go drinking in Glasgow and listen to Mozart or the one who doesn't want to max out his credit card?

The advocates became frustrated and started to assert their scientific expertise. When Stott argued that cosmic rays are affecting climate, Schmidt said, "This is completely bogus. You don't know that it's bogus, but I know that it's bogus." The skeptics, he complained, were "not doing this on a level playing field that the people here will understand." Thus, he managed to offend both his opponents and his audience, some of whom started shouting complaints and booing at this point.

I met Gavin Schmidt in his small, crowded office on New York's Upper West Side overlooking Broadway. Curiously, it is directly above Tom's Restaurant, scene of numerous *Seinfeld* episodes. The combative spirit of the show must have wafted up the airshaft because Schmidt turned out to be, without a doubt, by far the most argumentative person I interviewed for this book. I had not even reached the end of my first, rather prosaic question before Schmidt came back at me, challenging my terms and assumptions like a power attorney out to impress the jury.

Schmidt wished he had not gone on the NPR debate. "Looking back, I see that it's all about ego: assuming that obviously I'm right so I can prevail against somebody who is obviously wrong." He said he would not have done anything differently—except refusing to go on in the first place. "Political theater is not my game—it is never going to be a good venue for having a rational argument."

All this was still on my mind that evening as I idly surfed hundreds of cable channels and found John Stossel on Fox, complaining that no scientist would come and debate his resident climate skeptic. However, said Stossel, "we did find a scientist who was willing to talk about this so long as it was not a debate. So let's welcome NASA scientist Gavin Schmidt." Well, well, well, I thought—for all of his loathing of political theater,

Schmidt is prepared to appear on one of the most politically motivated programs on U.S. television.

Schmidt was excellent: clear, focused, and literate, presenting his science with the confidence that is such a crucial component in building communicator trust. Stossel demanded to know why he would not participate in the phony debate. "Because I'm not a politician," Schmidt said. "Anytime you want me to come on and discuss the science, give me a call and I will do that. But I am not interested in arguing with anybody just to make good TV."

Schmidt's uncompromising argumentativeness, which had seemed overbearing in his small office, here seemed entirely appropriate. A man who had been so wary of entering into the political debate was the only scientist prepared to walk straight into the most politicized venue on television, sit under a sign reading "GREEN TYRANNY," and explain his science on his own terms. Now maybe that needs a bit of attitude.

<u>29</u>

Two Billion Bystanders

How Live Earth Tried and Failed to Build a Movement

FROM HIS OFFICE IN BEVERLEY Hills, aptly named the Control Room, Kevin Wall organizes the largest public events around the globe: the World Cup opening ceremonies, rock concerts for Bob Dylan, Michael Jackson, Prince, and Elton John. He deals in mind-bogglingly big numbers. "Look," he tells me, "how many people saw Al Gore's documentary *An Inconvenient Truth*? Two million people, ten million people? That's a real success. It's the biggest documentary in history. I reckoned that I could create a twenty-four-hour show that touches on all the *Inconvenient Truth* stuff, and maybe one billion people, two billion people could see it."

And so Live Earth was born—simultaneous concerts in eleven cities, seven continents (including Antarctica), the largest and most ambitious attempt ever made to mobilize a mass audience around climate change.

Wall freely admits that when he first saw Al Gore's presentation, he had walked out halfway through because he found it so depressing. When the movie came out, his first thought was "Oh shit, it's the same slide-show." It did convert him to the issue, but he still had a big problem with the *slideshow*. Wall wanted to make an impact and that meant thinking big—big names, big venues, big aspirations, and putting his company on the line. "I might lose my money, my reputation, but I had an overriding passion and wanted to get that end result," he says.

155

Wall argues that the concert format enables an extraordinary access to the airwaves. "Try to get editorial control over three hours of prime time— impossible! But go to a network and get editorial control through a large entertainment charity concert that has your messaging in it? We can do that."

But, what was the message? Live Earth struggled to find a unifying narrative for climate change that could accurately present the issue while honoring the artistic integrity of the performers. They settled on the uncomfortable compromise of filleting short bursts of information and speeches in between the entertainment. In the event, this felt like a tax. As the British DJ Chris Moyles said after earnest filler on reducing your carbon footprint, "Serious stuff over! Shall we get back to the show?"

The usual green narratives were all in play. The grand finale at Wembley Stadium in London was preceded by a symbolic turning off of the lights. The lights came back up as Madonna bounded onto the stage in a black satin leotard, accompanied by a long line of children in school uniforms (sarcastically called the "Hogwarts School Choir" by the *New York Times*) and swore at the audience: "If you want to save the planet, let me see you jump. Come on motherfuckers, *JUMP!*" There were 150 complaints.

Her song, even if not that jumpable, started off upbeat and hopeful: "Hey you, don't you give up / it's not so bad / there's still a chance for us," before turning to a more naked self-interest in the second verse: "Hey you, save yourself / don't rely on anyone else / first love yourself." It was an unfortunate echo of the Live Aid anthem of a generation earlier to raise money for the starving Ethiopians, which had contained the killer line: "Tonight thank God it's them instead of you."

It was the images that ran on the video screens behind Madonna that grounded this firmly in the environmentalist version of climate change: smokestacks, traffic jams, flooding, hurricanes, starving dark-skinned children, chickens on conveyor belts, polar bears on icebergs, burning rainforest, melting glaciers, oil rigs, Martin Luther King Jr. The words of the song came up over the top like karaoke, bizarrely with the *s*'s transmuted into dollar signs: "ONE DAY IT WILL MAKE ENE."

One day it will make sense? Previous global events like Live Aid and Live 8, which Wall also organized, had made eminently good sense at the time. They had focused on bringing attention to issues that already had proximity: starvation in Ethiopia and debt alleviation. But Live Earth was building awareness for an issue that was still distant to most people.

Wall was "very disappointed" by the lack of follow-up on the momentum he had created, complaining that environmentalists had "done a lousy job of working together." Bob Geldof, organizer of the original Live Aid concert, had already warned that the absence of concrete measures meant that Live Earth would "just be an enormous pop concert." A spokesman for Live Earth responded to Geldof, saying that "people are aware of global warming but millions are not doing anything about changing their lifestyles."

And so the focus ended up on the minutiae of petty lifestyle changes, not on building a movement for change. The rock stars showed that they, too, were doing their bit. At the Tokyo show, the Japanese singer Ayaka told the stadium, "I started to carry my own eco-bag so I don't have to use plastic grocery bags, and use my own chopsticks instead of disposable ones." The hit eighties band Duran Duran opened its set with an ironic lifestyle tip: "Everyone who did not arrive on a private jet put your hands in the air," said lead singer Simon Le Bon, who also raised his hand. It contained an awkward echo of Michael Crichton's sarcastic comment in the live radio debate just four months earlier about his environmentalist friends flying their private jets to their second and third homes.

Maybe it all made ene to Madonna. A few weeks before the concert, she had discussed the environment over a macrobiotic dinner with the British prime minister in her eighteen-bedroom pied-à-terre in London. It's where she stays when she is not in her twelve-million-dollar house in Beverly Hills or her forty-million-dollar apartment in New York. A specialist contacted by the BBC estimated Madonna's annual emissions to be more than a thousand metric tons of carbon dioxide.

Climate change is a wicked problem in search of a narrative, and celebrities are not just people; they are narratives in their own right. The problem emerges when that celebrity-defined narrative (wealth, global fame, conspicuous consumption) comes into conflict with the environmental one (simplicity, locality, reduced consumption). As one viewer of Live Earth complained online: "Would you hold a hog roast to promote vegetarianism?"

Kalee Kreider, former communications director for Al Gore, has nothing but praise for the celebrities who participated in Live Earth, who, she says, put things on the radar that otherwise would not be there. "Does anybody say, 'Wow, that Princess Diana really brought *some baggage* to that land mines issue'? These people come to this as human beings doing the best they can with what they have. And you know what? God love them."

Ultimately, though, the celebrities were never going to be a measure of the event's success; they were the attraction that could bring in the punters to learn more, on the assumption that exposure to information would generate change. Gore told the media in the run-up to the concerts, "The tipping point in the political system will come when the majority of the people are armed with enough knowledge about the crisis and its solutions that they make this cause their own."

So this was a numbers game to arm enough people, and two billion viewers was a damned good stab at it. It was hoped that bringing so many people together would itself create the historic moment; as though the concerts alone could single-handedly create a social norm for action. But in the absence of a clear objective and a movement that could galvanize the audience into action, it created a global bystander effect: two billion people waiting on the sidelines to see if someone else would do something.

30

Postcard from Hopenhagen

How Climate Negotiations Keep Preparing for the Drama Yet to Come

GREAT HOPES WERE SET FOR the 2009 U.N. Climate Change Conference in Copenhagen—indeed, every subway station and bus stop carried the slogan "Hopenhagen," sponsored by Siemens or Coke. The city became a vast theme park for climate kitsch: fiberglass eco-globes by local artists, giant blow-up faces of indigenous people, melting ice sculptures of polar bears, a photo display of "100 places to remember before they disappear."

The Rådhuspladsen was dominated by a fifty-foot-wide illuminated globe on which energy saving tips alternated with the logos of the sponsors. On one side of Kongens Nytorv, there was a structure of scaffolding and flapping bedsheets, which on closer investigation I found to be scribbled with pleas for political action. And in Copenhagen harbor, alongside the statue of the Little Mermaid, the Danish sculptor Jens Galschiøt installed a new sculpture called *Survival of the Fattest*, featuring a grossly obese white goddess of justice (with a dinky set of scales) sitting on the shoulder of a frail African who was sinking under the waters.

The negotiations took place in a cramped conference center on the edge of town—accessible only by car or a little raised tramline that glided past shiny bland apartment buildings and a pointless Potemkin windmill erected by Siemens alongside the conference center. In the late afternoon light, it all looked unnervingly similar to the cool eco-world of Sustainia.

Inside the conference center, things were a lot less cool: ten thousand sweating observers and five thousand journalist milled about, holding meetings and press events for each other, taking photographs of themselves in front of the melting ice sculptures, or watching activists in polar bear costumes parading with placards reading "Save the Humans."

There was some grand talk in the political speeches, but it was mostly in the familiar slippery "we" form. Obama said, "We can act boldly, and decisively, in the face of this common threat." Wen Jiabao, the premier of China, said, "We will honor our word with real action."

The familiar metaphors clocked up. Todd Stern, the U.S. special envoy, called for "that old comic-book sensibility of uniting in the face of a common danger threatening the earth. It's not a meteor or a space invader, but the damage to our children will be just as great." In an opening address (reading from a script as full of exclamation points as a Valley Girl Facebook page), the Danish energy minister said, "This is our chance. If we miss it, it could take years before we get a new and better one. If ever! Let's open the door to the low-carbon age! Let's get it done! Now!"

And then they all watched the official opening film "Please Help the World"—four minutes of global eco-apocalypse in which a little girl desperately reaches for her stuffed polar bear cub toy as it falls down a crevasse in the cracking ground, just before the flood gets them both. Turn off the lights or the teddy bear gets it.

A year earlier, Yvo de Boer, the chair of the process, had warned that if this conference failed to set a deadline, "then one deadline after the other will not be met, and we become the little orchestra on the *Titanic*." It did fail to set the deadline. And the little orchestra did keep playing.

It is anathema to ever suggest that it might be better if the process should stop altogether. Oliver Tickell, the editor of the *Ecologist*, has struggled to get attention for proposals to reform the process. He says that the only thing that they really want to save is their annual meeting: "They negotiate overnight and then, sometime on Saturday, they all stand on the stage and say the same thing: We have achieved a historic breakthrough because we have agreed that by a date set sometime in the future we will agree on all outstanding points of disagreement. They may bicker, but they have a class interest in the continuation of the jamboree and to deliver just enough to keep it going."

And so President José Manuel Barroso of the European Commission duly admitted that he was disappointed, but that this was a positive step toward the many more steps in the future. U.N. Secretary-General Ban Ki-moon welcomed the Copenhagen conference as an "essential beginning."

The climate negotiations are always beginning, or in their favorite cliché, "setting the stage" for the drama to come. The U.N. declared that the Vienna climate talks in 2007 "set the stage for the Climate Change Conference in Bali." The U.S. Council on Foreign Relations explained that the 2009 Copenhagen conference "set the stage for ambitious action." According to Connie Hedegaard, the European commissioner for climate action, the Durban negotiations reached an agreement to "set the stage for the big deal in 2015."

"Setting the stage" is not just a diplomatic cliché—it is a narrative frame. It means that even when the meetings do not do anything, they are still preparing for the great drama to come. It is like the surreal films of Luis Buñuel in which the well-heeled dinner guests keep meeting for dinner but never actually get to eat.

This evasive circularity is now the rhetorical style of all official speeches about the process, with the same blocks of language reappearing and rearranging themselves in new patterns: "our goal must be nothing short of" "a call for a bold initiative" to "lay the foundations" with "concrete commitments" for a "roadmap for the further negotiations."

Or, as the British comedian Marcus Brigstocke put it in *The Now Show*'s "Dr. Seuss at Copenhagen":

> So they blew it, and wasted the greatest of chances,
> Instead they all frolicked in diplomat dances,
> And decided decisively, right there and then,
> The best way to solve it's to meet up again.
> And decide on a future that's greener and greater,
> Not with action right now, but with something else later.

31

Precedents and Presidents

How Climate Policy Lost the Plot

IT IS FORTUNATE THAT THE issue of climate change emerged at a time of unusual optimism, when there were three very recent precedents of proven success and international cooperation on which to draw. In retrospect, it is also extremely unfortunate that these issues had such strong metaphorical similarities to climate change that policy makers failed to notice the glaring and important differences.

The first was the Strategic Arms Reduction Treaty (START), negotiated between the USA and the USSR between 1982 and 1991. START established the use of targets and timetables in the cause of mutually verifiable reductions, which was carried directly over into greenhouse gas emissions reductions. Al Gore, then a senator, was a very active member of the bipartisan group of lawmakers that pressured President Ronald Reagan to moderate his stance in the START negotiations, and Gore refers to it regularly as a metaphor for decisive political action on climate change.

Gore, along with Tim Wirth, a Colorado Democrat, also led congressional efforts in the mid-1980s to mobilize U.S. support for a control on the production of ozone-depleting chemicals. This global struggle provided the second precedent and the mental model for the fight against climate change.

The ozone issue had first emerged in the mid-1970s, when satellites reported an extreme thinning of the stratospheric ozone layer over Antarctica. Like climate change, the invisible and otherwise harmless atmospheric gases that cause ozone depletion were the by-product of modern technology and lifestyles. Like climate change, they were damaging the planetary life support systems, especially at the poles. And like climate change, the only source of information was scientific experts and computer models, which were immediately championed by environmentalists and challenged by large corporations and libertarian think tanks. What is more, the main gases responsible for ozone depletion, chlorofluorocarbons (CFCs), are also powerful greenhouse gases.

Every one of these elements and players was remarkably similar to those of climate change and remarkably unlike those of any previous global threat. How could the two issues not become inextricably associated with each other?

Scientists learned the first lessons. In early 1992, ozone depletion over the Arctic turned out to be far less than NASA scientists had predicted. They were roasted by the conservative press for promoting a politicized environmental agenda. This chastening experience made the scientific community excessively cautious about making confident predictions on the climate issue that followed.

As the little girl in the TV commercial would like to know, looking up at Daddy, did the ozone story have a happy ending? Indeed it did, and that is what led it to exert such influence over the coming climate issue. The various actors played their prescribed roles—there was a David and Goliath struggle and a resolution that, as in all good myths, enabled the restoration of the status quo.

The model created by the ozone issue was one in which solutions lay in business-led technological innovation and the implementation of a market-based system of emissions permits enforced by binding international law: the 1987 Montreal Protocol. This is now regarded by the United Nations, which presided over the whole process, as "the most successful environmental protection agreement ever reached." Every single one of these elements was projected directly, and without challenge, onto the emerging issue of climate change.

So similar were the narratives of ozone and global warming that, for a long time, the public was thoroughly confused about which was which. A 1999 survey found that a quarter of Americans thought that ozone depletion was the main cause of global warming and, even thirty years after CFCs had been banned from aerosols, three-quarters of Americans still believed that spray cans caused global warming.

There was a third available precedent that promised a successful model for dealing with climate change: the success of market-based policies in reducing U.S. sulfur dioxide pollution. This, too, has uncanny similarities with climate change: gases (in this case sulfur dioxide and nitrogen oxides) that were the by-products of fossil fuel combustion were causing serious health problems and, in the form of acid rain, environmental damage across the United States.

Fred Krupp, the dynamic young director of the Environmental Defense Fund, had announced in a *Wall Street Journal* editorial the coming of a "third stage" for environmentalism that "respected growth, jobs, taxpayer and stockholder interests." Krupp duly championed a market-driven solution for acid rain pollution in which utilities could purchase and trade pollution permits on the competitive market. This model of emissions trading chimed perfectly with the free market conservatism of the Republican administration and was described as the "the holy grail of environmental policymaking." The reference to a mythical artifact is a strong hint that its appeal was in large part that of a culturally manufactured narrative.

Emissions trading embraced the free market as a means to reward innovation and protect powerful economic interests. Technology and engineering—in this case, scrubbers that would be applied to the smokestacks—could solve the problem. There was no need to abandon fossil fuels or constrain growth, and the demand for electricity grew by nearly a third over the next ten years. The problem was solved and the party could continue.

Five years later the world's major nations met in Berlin to discuss ways to meet their commitments to reduce their emissions of greenhouse gases. Following the Montreal Protocol, they looked to a binding international convention to be convened by the United Nations, this time to reduce greenhouse gas emissions. At the insistence of the United States (and with strong support from Vice President Gore), the approach to reduce these greenhouse gas emissions was transplanted directly from the acid rain legislation. Carbon would be given a market price by being converted into a tradable commodity so that countries could then swap their allowances. To this day, the U.N. declares that the "key tool for reducing emissions" will be the creation of the global market in carbon.

Carbon trading has become a deeply contested issue. Environmental activists adopted the phrase "carbon casino" and held demonstrations during the negotiations at which they threw photocopied banknotes into the air. Surprisingly, free market libertarians were equally disgusted, hearing in the words "market mechanisms" the code words for "market socialism."

Emissions trading also turned out to be a mechanism of byzantine complexity that distributed responsibility and severed the connection between personal behavior and moral responsibility. With trading, it made no difference whether you flew or drove or whether you bought electricity from a wind farm or a coal plant—the emissions had already been decided and allocated as permits.

Nor did it even work. The over-allocation of permits and a flood of fraudulent Russian and Ukrainian gas offsets have led, in the words of leading analyst firm Thomson Reuters Point Carbon, to a "market melt-down" in the Europe trading scheme. By 2013 polluters had banked so many cut-price permits that they could expand emissions enough to outweigh the savings of all the European renewable and energy efficiency efforts combined.

In 2010 with a new elected president sympathetic to climate change issues, a concerted push was made to get climate change legislation within the United States. Once again Fred Krupp from the Environmental Defense Fund played a leading role and once again emission trading was presented as the only viable vehicle for reducing emissions.

As was found in the U.N. and Europe, the supposedly simple and efficient market mechanisms required a vast and verbose technical manual. When it finally limped out of the House Energy and Commerce Committee the American Clean Energy and Security bill had blown up into a 1,428-page epic of monitoring, evaluation, and allocation processes. Even with very generous allocations for large oil and coal interests to buy off their resistance, the bill was doomed to fail.

At an international level, the U.N. designed the Clean Development Mechanism (CDM) for projects in developing countries to trade their emissions savings with northern polluters. It was described by the *Economist* magazine as a "shambles" with widespread allegations of fraud. More than half of its alleged savings came from a handful of Asian companies that were producing an obscure and very powerful greenhouse gas called HFC-23 largely so that they could claim credits for destroying it.

In 2012 the CDM executive board announced that credits could be awarded to coal-fired power plants in the developing world if they improved their efficiency. Now a new coal-fired power plant in Europe can "offset" its emissions by buying carbon credits from another new coal-fired power plant in India. This seems to be not so much robbing Peter to pay Paul as robbing everyone to pay them both off.

Reviewing this sorry history, academics Steve Rayner at Oxford University and Gwyn Prins at the London School of Economics concluded that arms reduction, ozone depletion, and sulfur emissions should never have been chosen as models for action on climate change in the first place. These were, they said, "tame" problems with well defined and achievable ends. Climate change, though, is a "wicked" problem of altogether more daunting scale, complexity, and uncertainty. "Experience," they said, "can carry fatal baggage."

Fatal indeed. There was cognitive error on a vast scale that, tragically, perfectly sums up all the flawed psychological processes discussed in this book. Decision makers and policy strategists are no different from anyone else and are bound by the same cognitive limitations. At each stage, they drew on the readily available precedents and made assumptions based on simplistic and metaphorical similarities. Through ever-more-energetic confirmation bias, they promoted a social norm to their peers that led them to keep repeating the same mistakes.

These precedents were on an entirely more manageable scale than climate change. The number of actors involved in both issues was very small—a mere 25 power utilities and 110 plants were involved in the Acid Rain program. Twelve companies and their subsidiaries accounted for the vast majority of the production of ozone-depleting chemicals—a quarter of global production was by DuPont alone. The CEOs of all these companies could comfortably have attended the same cocktail party.

What is more, the damage caused by ozone depletion and acid rain could, after the pollution was controlled, be reversed within a generation. These issues created an optimistic narrative of resolution and renewal that was entirely inappropriate for the irreversible and open-ended problem of climate change.

Frames do not just focus the attention: they define the areas for *disattention*. These precedents bound climate change to a limited set of meanings that actively excluded other approaches. They defined climate change as an environmental issue and therefore not a resource, an energy, an economic, a health, or a social rights issue. They determined that it would be best managed through emissions trading, and therefore not through regulation, taxation, and rationing. And the U.N., glowing from the success of its process to prevent ozone depletion, determined that climate change would be best controlled through international protocol rather than regional or multilateral agreements.

But the largest, most extraordinary, and damaging misframing of all acquired from the precedents of ozone depletion and acid rain was that climate change could be defined entirely and exclusively as a problem of *gases*.

This may well prove to be our fatal mistake.

32

Wellhead and Tailpipe

Why We Keep Fueling the Fire
We Want to Put Out

IN SOME WAYS, CLIMATE CHANGE is actually quite simple. We find fossil fuels. We dig and pump them out of the ground. We process them and sell them. Then we burn them. The waste gases include carbon dioxide, which traps the heat in the atmosphere that leads to global warming. There is, of course, much more to it than this—other gases, sources, and sinks—but this is the basic carbon cycle, which represents the majority of the problem and appears in every textbook and informs every policy.

So, there is a chain, or, if you prefer, a pipeline. At the one end is exploration, development, and production—what I will call the *wellhead* (a term which will include the minehead). And at the other end is the sale and then combustion that leads to emissions—what I will call the *tailpipe*. Policies to manage climate change should, one would think, consider interventions at both ends and all stages in-between.

However, they do not. The focus on tailpipe gases and disregard for wellhead fuels has been the single most important factor in all government and policy framings. Radical environmentalists alone have attempted to connect two issues that, in the minds of most mainstream experts, operate in entirely independent realms. This does not, on its own, explain why we ignore the risk of climate change, but it does explain the fundamental disconnection that works through all narratives and policies on the issue.

It explains how a climate science funding body can subsidize exploration by oil and gas companies, and how the Science Museum in London can have a display on climate change funded by Shell. It explains how Norway, the world's eight largest exporter of oil, can see itself as a champion of action on climate change. It explains how President Obama, speaking as usual in the first-person plural, can boast in a 2012 presidential debate that "even as we're producing more coal we're producing it cleaner and smarter, same thing with oil, same thing with natural gas" and then, just a few months later, say, "We need to act, but we can't just drill our way out of the energy and climate challenge."

It explains how Hillary Clinton, who calls the climate crisis "the chief threat of the 21st century," could visit Norway in June 2012 to negotiate U.S. access to the nine hundred trillion dollars in Arctic oil reserves. Her internal dissonance came to a head on June 2. She started the day with a trip on a scientific research vessel to see the melting Arctic—an experience she described as "sobering." Back on land, after a lunch of local seafood delicacies, she went straight into a roundtable attended by the CEO of Norway Statoil and the country director of ExxonMobil to plan the expansion of Arctic oil production.

In Britain, energy and climate change are combined into one government department leading to simultaneous action to reduce emissions *and* to boost oil production. One month the Minister of *Energy* and Climate Change brags about the allocations of new licenses to release twenty billion barrels of oil around British coasts. The next month the Minister of Energy and *Climate Change* announces an ambitious plan for the government to reduce its emissions by 10 percent.

Although many campaigners would regard such inconsistency as evidence of the hidden influence of oil corporations, policies to deal with other global problems—even those with powerful vested interests—never ignore production in this way. Fisheries are managed through fishing rights and production quotas. Illegal logging is prevented through permits and forest management. And to take a notoriously wicked problem, it would be unthinkable for drug policy to ignore production, which is why the U.S. government spends nearly two billion dollars per year on international control measures. If, as George W. Bush said, we are addicted to oil, then a policy on climate change that ignores production of fossil fuels is like a policy on drugs that ignores the poppy fields, cocaine labs, smuggling networks, and dealers and focuses exclusively on the addicts.

In fact, when compared with these other examples, it makes even *more* sense for climate change policies to include production. Two-thirds of the world's oil production is produced by just ten oil companies or, to look at it another way, by just ten countries. Representatives from every one of them could comfortably sit around a single conference table. There would be enough space left over for a couple of extra delegates from China and the United States, which between them account for two-thirds of the world's coal production. This was self-evident to the economist Thomas Schelling, who won the Nobel Prize for his work on negotiation theory. Schelling regards climate change policy to be an "awfully complicated hodgepodge." To him, the answer seems very straightforward: "The way to simplify this is to put the cap on the fossil fuels, not on different industries—a cap on oil and gas at the wellhead, a cap on coal at the minehead," he says. When I asked him why he thought that no one has ever proposed doing this, he said, "I don't have any good theory except to say that hardly any leading politician in the U.S. manages to speak truthfully and candidly about the subject of climate change."

Robert N. Stavins, a professor of business and government at Harvard University, agrees. He supports the model of emissions trading but argues that the precedent should have been the trading model applied during the 1980s to phase out leaded gasoline, which focused on the lead content in gasoline (the fuel input) not the exhaust (the tailpipe output). According to Stavins, the cheapest and most enforceable system is to tax the carbon content of the fuels.

So, I wondered, when had the governments and experts working on the international process weighed up the various options for policy interventions and decided that the best basis for national and international policy was to regulate and trade gases and ignore the fossil fuels that produce them?

The answer is that this discussion has never taken place.

The science of climate change has only ever been concerned with greenhouse gases and their potential impacts. In 1990 the Intergovernmental Panel on Climate Change (IPCC) produced the first tables of national greenhouse gas emissions, which then became the basis of the international negotiations. However the IPCC had no remit to produce tables of national fossil fuel production. Sir John Houghton, the founding chair of the IPCC, sees no difference between the fossil fuels that are produced and the greenhouse gases they later become. "Of course," he says, "they are all part of the same thing." The problem, he told me, is that "talking about the source of fossil fuels would have moved us from the science arena into the policy arena. Because of the pressure we were under, we needed to be squeaky-clean, maybe too clean, but we needed it to be that way."

In his fourteen years chairing the IPCC Science Committee, Sir John could not recall a single proposal or debate about controlling production at the wellhead. "It's a pity that it never has been addressed, but it is not a science question. It is a policy question."

Except that it was never discussed in policy circles either. Jeremy Leggett, a former oil industry geologist who later became the senior climate campaigner for Greenpeace International, has attended the international climate negotiations since 1990 and sat in on every major policy discussion. Yet he could not recall a single instance when anyone formally proposed a policy that would constrain the exploration and development of new sources of oil, gas, and coal. "Looking back," he says, "I think it was a missed opportunity that we did not make this argument earlier."

"Nor," he adds, "was there any point at which the oil companies or producers actively generated this approach. It just happened to be the zeitgeist that this was how the problem was approached and this just happened to be in the interest of the oil companies."

Leggett's observation is important. Of course, there were very powerful national and commercial interests protecting fossil fuel production. Attempts to limit wellhead production would have been bitterly resisted. But there were no fights, no struggles, no backroom deals. There did not need to be because it was never discussed.

Jennifer Morgan has dedicated most of her working life to following the climate negotiations. She headed the delegation for the World Wildlife Fund throughout the Kyoto negotiations and has led campaign teams in every negotiation since 1994. Like Leggett, she could not identify any point at which a decision was made to focus on tailpipe emissions. She tells me that "it has just been assumed for a very long time." Morgan and Leggett agree that campaigners and negotiators alike were trying to get the best deal they could on the terms that were available to them—and these were only concerned with the tailpipe.

And so everyone fought bitterly over the mechanisms for sharing, controlling, and trading tailpipe emissions. The science was only concerned with *gases*. The international process had taken all of its precedents from previous policies to control and trade *gases*. From the very outset fossil fuel production lay outside the frame of the discussions and, as with other forms of socially constructed silence, the social norms among the negotiators and policy specialists kept it that way.

Consequently, the narratives surrounding the environmental impacts of fossil fuels became divorced and disconnected. Narratives around the impacts of oil, gas, and coal production become concerned with health, safety, and compensation for localized environmental damage. Narratives around climate change become concerned with emissions, energy demand, efficiency, consumer lifestyles, and global climate impacts.

In spring 2010 there was a perfect opportunity to bring these two drifting stories back together again. Senators John Kerry and Barbara Boxer were holding intensive negotiations to redesign a climate bill for a vote in the U.S. Senate. To maintain the political momentum, mainstream environment organizations were running the largest and most expensive public outreach campaign ever held on climate change. And then, on April 20, BP's *Deepwater Horizon* oil rig exploded and spilled four million barrels of crude oil into the Gulf of Mexico.

The result should have been a godsend for climate advocates: a media-driven outrage about dirty fossil fuels at exactly the time that legislators were considering an even greater crisis caused by the pollution of those same fuels. This perfect storm of synchronicity happens only once in a campaigner's lifetime.

Only there was no room for making these connections because the U.S. climate bill, following the lead of the international negotiations and the acid rain emissions trading, was entirely concerned with tailpipe emissions. In return for their involvement and support, oil, gas, and coal companies had been offered generous emissions allowances and increased exploration rights. Three weeks before the explosion, as part of a grand bargain to get a climate bill through the Senate, President Obama had announced plans to open up 167 million acres of the Atlantic coast for oil drilling.

So there was no means to connect the narratives—although grassroots groups like the Sierra Club tried valiantly to do so. *Deepwater Horizon* became all about local impacts and corporate health and safety—with a compelling enemy narrative centering on the slippery and evasive performance of BP CEO Tony Hayward.

The more radical environmentalists have always tried to make the connection between wellhead and tailpipe. Patrick Reinsborough of the Oakland-based Center for Story-based Strategy sees this as part of a wider vision of an inclusive movement for social justice. "There is now a new political space opening up," he told me, "in which we see that the problems are fossil fuels, and people who have been on the front lines of fighting fossil fuels are the obvious people who should be on the front line fighting for solutions."

The battle over the Keystone XL pipeline and the movement mobilized to demand divestment from fossil fuel companies are also determined attempts to reframe the issue of climate change in terms of wellhead. A pipeline not only has the proximity of a visible struggle but it also gives tangible form to the transmission of carbon from stored reserves to emissions.

But the problem remains—as with the language and images of climate change as a whole—that for as long as radical activists are the only ones making these connections, their arguments may be marginalized and disregarded.

So, I suggest, the separation of wellhead and tailpipe was not primarily the product of corporate lobbying or global power politics—although these have played a role. It can also be understood as an extreme error of judgment resulting from cognitive error and flawed categorization. Scientists categorized climate change as a tailpipe issue because production was considered a political issue that was outside of their domain. Policy makers then categorized climate change as a tailpipe problem because they drew on recent available experience that suggested viable solutions for tailpipe problems. Confirmation bias and a socially constructed norm of disattention finished off the job.

Because climate change is multivalent and wicked, it can have multiple interpretations but exists only in the form that people choose it to have. This means in turn that it does not exist in the form that they choose to ignore. Production controls are not debated, because they simply don't exist within the debate.

After twenty years of negotiating around emissions, we are now in a bizarre situation. Most Western governments have established programs to subsidize the increasing production of renewable energy, biofuels, and—with less success—nuclear power. And they do so while encouraging, and usually subsidizing, ever-larger investments into exploring and developing new fossil fuels.

In 2012, the global investment in renewable energy was an impressive $244 billion. In 2012, for the first time, the oil and gas sector investment into exploration and development of new reserves broke the $1 trillion barrier. While the renewable sector tries to reduce emissions at the end of the pipe, the oil and gas industries seek to pump ever more fossil fuels into the front of the pipe.

But the procession must go on and the lords of the bedchamber are taking greater pains than ever to appear to be holding up the emperor's train, although, in reality, there is no train to hold.

33

The Black Gooey Stuff

Why Oil Companies Await Our Permission to Go Out of Business

SHELL OIL IS VERY CONCERNED about my safety. On a large video screen by reception in its South Bank headquarters in London, the office manager tells visitors, "Your safety is very important to us. We have a policy of Goal Zero: *All* accidents *can* be avoided. Comply with all safety notices. Do not run in the corridors. Report any unsafe behavior by your host. We hope you have a very happy, positive, and *safe* visit."

As I am watching this, the woman at reception is trying to get my attention.

"Your shoe," she says.

"Sorry?"

"Your shoe. Your shoelace. *Shoelace.*" By now, everyone in reception is joining in and pointing. "Your shoelace, *sir*. It's loose. You could trip up and have an *accident.*" "Oh well," I say, "I like to live dangerously."

Following the circular staircase that spirals up to the meeting rooms are cardholders, like the kind they have in fancy cake shops, with smiling faces of the Shell family. It's very exciting—say the people on the cards. Never a dull moment. Vibrant. Exciting. We have a shared vision. We know what we want to be. The head of communications says, "One big challenge is to help Shell show people in the UK that it's really serious about climate change." A big challenge indeed—but then Shell likes challenges.

175

As I make my way up, reading the cards as I go, the receptionists are concerned again. Now they are sounding more anxious and obviously have me pegged as a dangerous risk taker and threat to Goal Zero.

"Sir. *Sir!* Can you *please* hold the handrail on the stair."

"The what?"

"The handrail—you might trip and have an *accident.*"

It is then that I notice an array of signs all along the staircase warning me of slippery steps and directing me to walk carefully and use the handrail. When I finally sit down with David Hone, Shell's climate change supremo, a look of concern crosses his face.

"Your pen," he says.

"Pardon?"

"Your pen, George. Your pen. It's leaking. It could spill and mess up your clothes."

And I want to say, "Your oil, sir. Your *oil.* It could burn and destroy the biosphere." Seeing that I have been invited to "report any unsafe behavior by your host," there is a lot more that I could have said about the safety of selling products that dump 380 million metric tons of carbon dioxide into the atmosphere each year.

This obsessive focus on personal safety would be odd under any circumstances. The fact that it is so prominent within an institution involved in the most dangerous activity in human history suggests that a major oil company might be just as prone to developing irrational avoidance mechanisms and bizarre self-justifying narratives as any individual or social group. And why not? It is, after all, a social network with its own identity, internal culture, and social norms.

By this reckoning, is it surely relevant that Shell, like all the oil giants, deals with climate change within a single combined Health, Safety and Environment department.

Professor John Adams of University College London howls with laughter when I later tell him the Shell story. Adams is the author of the most popular textbook on the social construction of risk and has a particular loathing for health and safety policy, which he calls the "compulsive risk assessment psychosis" (a.k.a. CRAP). He recalls that when he gave a presentation at MI6, the secret intelligence service that supposedly employs James Bond, a hardened security operative complained bitterly about being chastised for unsafe stair walking in the head office.

Adams's academic research concentrates on the role that our

perception of control has on our perception of risk. He observes that the excessive focus on health and safety is common across all resource extraction industries and derives from a culture that sees risks as something that can be managed and controlled. There is, he agrees, a bias derived from definitions of influence and control behind Shell's concerns about safety. "I see cultural bias everywhere I turn," he says. "My own cultural bias is to see the world though cultural bias theory." I guess that's a meta-bias.

Safely seated with Hone, fortified by petits fours from the cornucopian nibbles buffet that fuels Shell's power meetings, familiar cultural biases weave their way through our conversation: in groups, out groups, social norms, bystander apathy, self-justifying enemy narratives, and, as usual, the total separation between wellhead and tailpipe.

Hone is a straight-talking Australian chemical engineer who has worked for Shell all of his life. Like the people shown on the cards on the stairs, he finds it very *exciting* to be working in such a powerful company. He has just returned from Alberta, Canada, and talks with awe of Shell's grand ambition for its operations in the tar sands. Once again, I am reminded that one person's environmental disaster is another person's engineering achievement.

Hone is fiercely loyal despite having the unenviable task of defending Shell's role in an issue that he openly admits is an extremely "inconvenient truth" for his company. I am pleasantly surprised, having expected to hear a flow of carefully crafted PR lines, to find that Hone comes across as a decent and honest person. "It's a bugger," he says.

Back in 2009, he was filmed talking of his conviction, as a father, that it is a "no-brainer that we move to a new carbon pathway" and how excited (that word again) he was about the transformation under way in his company. A few weeks later, Shell froze all of its investment in renewable energy.

The climate activist Bill McKibben says, "This is, at bottom, a moral issue; we have met the enemy and they is Shell." Well, I'm all up for meeting enemies and understanding how they operate. So I am intrigued to know how Hone squares his concerns about climate change with his loyalty to a company that apparently has no brain and has decided to spend ever greater amounts—currently more than thirty billion dollars a year—bringing more oil and gas reserves into production.

Hone's argument, not surprisingly, adheres strongly to the tailpipe

narrative: that the responsibility lies with the emitters who give Shell the
"permission" to extract the fossil fuels that they choose to burn. *Permission*
is a key word for Hone, and it crops up throughout the interview. "We
need the permission that society gives to us," he says, but the oil industry
"is not being given the permission to make a transition out of fossil
fuels." And the main reason for this is that "the international agenda is
driven by people with political agendas that are unrelated to solving the
problem."

These distracting agendas include development, social rights, and
poverty campaigns, but especially environmentalists who, he says, have
written climate change as the old socialist-versus-capitalist fight in a
different format.

Hone wants it known that this familiar enemy narrative is his personal
view rather than that of his company. Really though, it is just a blunter Aussie
version of the language that permeates the future scenarios produced within
Shell. The most recent of these warns of a future dystopia in which "powerful
climate lobbies" demand "disruptively overreactive, ill-considered, politically
driven knee-jerk responses"—including, shockingly, bans on the development of new sources of fossil fuels. This, by some complex self-serving logic,
leads the tailpipe emissions to grow "relentlessly."

Shell is understandably nervous about any restriction on its expansion.
The British nonprofit organization Carbon Tracker Initiative has been
warning institutional investors that the current carbon reduction targets
(inadequate though they may be) cannot be met without leaving 60 to 80
percent of the currently listed oil and gas reserves in the ground. This in
turn means that the share value of Shell, which is underpinned by these
reserves, is grossly inflated. It is hardly surprising that Shell executives
prefer to keep their eyes focused on the risks of slippery floors.

So climate change is a collective bugger of an inconvenient challenge.
Shell wants to keep finding the stuff and digging it up. We keep burning
ever more of it because we are addicted to energy. And everyone is threatened by climate change. The solution to this quandary, which, as usual,
Shell is not being given the *permission* to develop, is carbon capture and
storage (CCS).

CCS is a group of technologies that can remove carbon dioxide from the
waste flue gases (or in the case of gas, from the fuel itself) and then pump it
into underground aquifers for permanent storage. Cynics suggest that all that
dangerous carbon had been quite safely stored underground to begin with.

There are currently eight large-scale CCS projects and eight more under construction, which, between them, will soon be storing thirty-six million metric tons of carbon dioxide a year. This sounds promising until one considers that we will need sixteen thousand more plants on this scale to deal with current emissions. Emissions are still increasing so fast that we will need another thousand plants each year just to keep up with the annual increase. And transport emissions—which are growing even faster—cannot be captured by any technology.

Legitimate debates continue about whether these technologies could ever be economically viable, whether the carbon dioxide could really be stored permanently, and whether there is enough space to put it. The costs of CCS are extremely uncertain, running at $150 per metric ton of CO_2, and the technology becomes viable only if CO_2 can be captured at a cost of $25 per metric ton. There is a long way to go.

David Reiner, an expert in CCS at Cambridge University, is convinced that it is technically possible, providing that there is economic pressure and political will. He tells me that CCS has no economic justification on its own: It is entirely linked with concern about climate change and, he says, "CCS development moves forward only when there is a lot of interest or concern in climate change."

Of course it does. CCS is the card that gets everyone out of jail. CCS allows governments to carry on talking big on climate change while continuing to expand oil and gas production. It provides the reason that oil and gas companies can stay in business. And it fits perfectly into the industry worldview of engineering solutions. As ExxonMobil CEO Rex Tillerson says, "Maybe I'm biased because I'm an engineer, but I have enormous faith in our technology's ability to find solutions as they present themselves to us." Tillerson named it correctly: There is an explicit bias at work.

Maybe CCS can work. I hope it does—this is one engineering technology that could make a genuine contribution to the transition out of fossil fuels. My fear, though, is that CCS is less of a real solution than a much-needed narrative ploy, providing the vital missing piece in a story that is unstable and incomplete without it. For narrative purposes, CCS does not need to work on a large scale. It doesn't need to be economically viable or ever be competitive with renewable energy. It scarcely needs to work at all. All that is required is that it exist tangibly enough to provide an alibi for the wellhead—a few demonstration sites, some chunky reports. And then lots of creative storytelling about human ingenuity.

Shell, for example, says that CCS is, along with nuclear, the "hinge" on which future emissions depend. BP says that CCS is a "critical technology for reducing emissions." A high-level working group on the future of coal at the Massachusetts Institute of Technology in 2007 also concluded that CCS is the "critical enabling technology." Shell does not anticipate large-scale deployment of CCS until 2030, but then, within twenty years, it predicts that the developed world will convert 90 percent of coal- and gas-fired plants to CCS. There is some textbook temporal discounting and optimism bias going on here.

"Never forget," says Steve Kretzmann, founder of the environmental campaign group Oil Change International, "never, *ever* forget that the oil industry is the most extraordinary wealth-generating machine ever invented by man. That is what it is designed to do and it does that very well. I haven't seen a single intelligent thought about how you could transition the industry out of oil and keep it just as profitable."

Kretzmann has spent twenty years campaigning against oil companies and has watched the same cycle over and over again: "Every so often when they think something is going to happen, they buy a whole load of wind, sun, algae, CCS, whatever. Then their lobbying arms do their jobs, the threat recedes, and they say, 'Nah, we're not going to do this.'"

The journalist Ross Gelbspan interviewed six U.S. energy and oil company presidents for his book *The Heat Is On*. All but one of them agreed that climate change was happening. Like David Hone, they all said the same thing: that as soon as governments regulate climate change, they would become "energy companies." In the meantime, says Gelbspan, they admitted, off the record, that the competitive environment forced them to suppress the truth about climate change and ensure that those regulations do not happen.

Gelbspan tells me he has "pondered a great deal how these men could be loving grandfathers and such cold-minded executives." His only conclusion is that they are outstandingly good at compartmentalizing different areas of their lives and preventing any connections from jumping across those boundaries.

Kretzmann is surprisingly empathetic when he talks about oil company employees and recognizes that many are struggling with these contradictions and "genuinely believe that their chance of changing things is better on the inside."

That being said, when it comes to the battle for public opinion, it's gloves off. One of the TV commercials produced by Oil Change International contains smiling "executives" chatting to the camera in the intimate way currently favored by corporate advertising: "Begging a fortune in subsidies, destroying your kids' future. At Exxon, that's what we call good business. Here at Exxon we hate your children."

"They don't like it much," Kretzmann says, and chuckles. He makes no bones that the public fight is not really about policies and climate data; it's about identifying enemies and reframing oil. "Whenever the oil industry wants to talk about their great benefits, they talk about energy. When we want to talk about their impacts, we talk about oil—that black gooey stuff."

34

Moral Imperatives

How We Diffuse Responsibility for Climate Change

WHEN DICK CHENEY, FORMER U.S. vice president, accidentally shot his friend Harry Whittington in the face, he drew on his forty years in politics and the oil industry to create four stages to the process. "Ultimately," he said, "I'm the guy who pulled the trigger that fired the round that hit Harry." His boss, George W. Bush, added yet further levels of detachment—"He heard a bird flush, and he turned and pulled the trigger, and saw his friend get wounded."

Even when they shoot their friends, it seems, politicians will do everything they can to create multiple stages between themselves and responsibility. It is a peculiarity of English that it doesn't distinguish between intentional events and accidental events. In many languages, you would say "I broke my arm" only if you went mad and broke it. But English makes up for this weakness with the ingenious passive voice, which removes intentionality: My arm was broken; my friend got wounded; the climate got fried.

And so it is with climate change. We diffuse the responsibility into multiple stages, each one protected by the passive voice. The oil is produced. It is burned in the car. The climate is changed. Someone's life is destroyed in a climate disaster on TV. Or, as the *Wall Street Journal* says, "climate concerns cannot and must not be

182

ignored"—this, incidentally, in a quotation from a report arguing for the production of Canadian tar sands.

So where in this chain *does* responsibility for climate change lie? Stephen Gardiner, professor of philosophy at the University of Washington, argues that *all* the decisions concerning action or inaction on climate change are ethical issues—especially in regard to intergenerational rights.

Which, for Gardiner, raises an interesting question: Why is there so little discussion of the ethics of climate change? The answer is that this is yet another area in which people shape the issue to avoid the discussion. Certainly no one is in a hurry to invite ethicists into the policy discussions. In 2010, the United Nations considered creating a Universal Declaration of Ethical Principles in Relation to Climate Change. After ten international consultation meetings, its ad hoc working group concluded that the U.N. "should be given the opportunity to review the desirability of preparing a draft declaration." I don't think that was a yes, even by U.N. standards.

This is exactly the kind of obfuscation that prompts Gardiner to suggest what he calls "a very unpleasant thought" that politicians deliberately create needlessly complex treaties and unworkable processes to draw attention away from the need to do something. This observation is not lost on other commentators. *Guardian* journalist George Monbiot argues that "government policy is not contained within the reports and reviews it commissions; government policy *is* the reports and reviews." He says, "Government creates the impression that something is being done, while simultaneously preventing anything from happening."

The key factor that determines moral responsibility is intentionality. Humans are acutely alert to interpreting people's intentions—even children as young as three respond differently to identical harmful acts depending on whether they regard them as intentional or not intentional.

The reason that an enemy narrative motivates the emotional brain is because an enemy has the clear *intention* to *harm us.* If scientists had discovered that North Korea was pumping greenhouse gases into the atmosphere with the intent to destabilize the world's climate, there would be immediate political consensus to take action, regardless of the cost. That, of course, would be a very big problem but, crucially, a *tame* problem, and far more easily solved.

So climate change struggles with intentionality. No one wanted climate change to occur. No one ever purposefully wanted to hurt anyone through climate change. As the journalist Gwynne Dyer argues in his book *Climate Wars*: "Nobody is to blame for the crisis that hangs over us—not my mother who had five children, not William Levitt who invented the modern suburb, or Henry Ford."

But that is in the past. It is much harder to argue one's innocence when one *knows* that one's actions are causing harm. If climate change becomes intentionally harmful only when people *know* they are causing it, is it any surprise that most people do everything they can to avoid learning about it or accepting that it exists?

Like many skeptics, U.C. Berkeley physicist Richard Muller would be very happy if there was no discussion about morality at all. Talk of responsibility, he says, is all about blame. Muller recalls a Frenchman accusing Americans of being arrogant for blaming themselves—as though only America could be important enough to count. We both have a laugh at this: Is blaming someone for blaming themselves some kind of meta-blame, I wondered. This issue, Muller says, needs "problem solvers not blame seekers."

But if we want to be problem solvers, we still have to decide exactly where that problem lies. Whether we are concerned with wellhead, tailpipe, or both, we still have to agree who is going to make the changes. And that leads straight into the issue of fairness and back into the ethical dimension. Ethics are unavoidable.

Everyone is strongly in favor of the principle of fairness. The problem is that everyone also defines fairness in terms of his or her own self-interest. This can be taken to a ludicrous extreme. The Republican chairman of the House of Representatives debate on the Kyoto Protocol, David M. McIntosh, argued that the protocol was "patently unfair" because it exempted countries that already had the "competitive advantages of cheap labor, lower production costs, and lower environmental, health, and safety standards." Such, it would seem, are the unfair economic advantages of grinding poverty.

Psychology research suggests two key reasons why it is proving so hard to define fair reductions. The first is that our attachment to the status quo leads us to give an excessive value to what we already possess. We come to believe that this originates in our own skill, talent, and hard work and is therefore a fair reward.

The second reason is that while people are sensitive to losses, they are even more sensitive to the fair distribution of losses. In experiments, people may tolerate an unfair distribution of gains in the interest of a quick settlement, but they will doggedly insist that any loss is unfair, even if, by delaying an agreement, they end up paying far more.

This problem constantly recurs around the management of shared environmental resources where everyone wants the gain of exploitation but no one wants to accept the loss of constraint. In his highly influential and highly disputed paper published in *Science* in 1968, the ecologist Garrett Hardin argued that we are all forced by our evolutionary drives to maximize our personal benefit from a common resource even when we know that this will lead to its ultimate destruction. He named this phenomenon the "tragedy of the commons."

Not surprisingly, climate change has been called the *ultimate* tragedy of the global commons, although, as usual, the people who use the phrase invariably focus on the tailpipe-emissions-into-the-atmosphere commons rather than exploitation of the common reserves of fossil fuels.

Such is its fame that people tend to forget that Hardin's paper is not a reasoned argument grounded in evidence but an ideological polemic grounded in prejudice. Its primary aim was to confront "liberal taboos" in order to argue that the provisions of a welfare state encourage "overbreeding" by the poor. None of this concern about overpopulation prevented Hardin from pursuing his own hard-wired self-interest and having four children of his own.

Hardin's deterministic model of human nature melds perfectly with the interests of authoritarianism and economic elites. Thus, when speaking of the atmospheric commons, Hardin says, "The air and waters surrounding us cannot readily be fenced, and so the tragedy of the commons as a cesspool must be prevented by different means, by coercive laws or taxing devices that make it cheaper for the polluter to treat his pollutants than to discharge them untreated."

If climate change is a tragedy of the commons, it follows that appeals to responsibility and conscience are a waste of time and that, in Hardin's words, only "mutual coercion mutually agreed on" will work to curtail our insatiable personal interests.

But there are many other ways to see it. The political scientist Elinor Ostrom won a Nobel Prize in Economics for her research into the innumerable ways that people collectively manage resources. In a direct challenge to Hardin, she argued that people will sustain and even improve shared resources providing there is free communication, a shared vision, a high level of trust, and a mobilization of participating communities from the bottom up.

And if this is, as Stephen Gardiner puts it, an issue of collective moral pollution in which we benefit ourselves at the expense of future generations, then we need to build that bottom up vision by agreeing to a set of principles based on our shared values. The problem is that we do everything we can to avoid thinking about climate change in any form, including, as we shall see, its implications for our own children.

35

What Did You Do in the Great Climate War, Daddy?

Why We Don't Really Care What Our Children Think

IN 1915 ARTHUR GUNN, A London printer, was debating whether to join the army. He said to his wife, "If I don't join the forces, whatever will I say to Paul if he turns round to me and says: 'What did you do in the Great War, Daddy?'" Gunn suddenly realized he had a marvelous slogan for a recruiting poster and passed a sketch onto a propaganda artist he had worked with, Savile Lumley. Lumley did a fine job, choosing to make the interrogator a little girl. Sitting on her father's lap, a history book on her knee, she asks him this probing question as he stares wistfully into the middle distance.

The moral dilemma that inspired this iconic poster is one of the recurring ethical themes in climate change communication. Caring for the welfare of our children is one of our strongest evolutionary drives and one of the few concerns that consistently overcomes self-interest. On the face of it, giving those children a voice in our decisions, especially imagining how they might confront us in the future, should be a powerful spur to action.

The veteran environmentalist Jørgen Randers, one of the authors of the famous 1972 *Limits to Growth* report, tells us that our "first priority should be to prepare the foundations for an unassailable answer to the

question, 'What did you do, (grand)father, when greenhouse gas emissions were allowed to grow out of control in the early 2000s?'"

Al Gore invites audiences to consider which question they want to
hear from future generations: "What were you thinking? Didn't you see
the North Pole melting before your eyes?" or "How did you find the moral
courage to solve the crisis?"

This narrative applies to businesspeople too. John Varley, the former
chief executive of the international Barclays Bank group, says, "More
than anything, I want my children to be able to look me in the eye and to
say with conviction 'You played your part.'" Hopefully his children might
also spend a few minutes on the Internet and then ask him why he
approved nearly six billion dollars in loans to companies mining or generating power from coal.

It not surprising that President Obama, as a committed father, speaks
often of the welfare of children and being able to reply proudly to them
that "this was the moment when the rise of the oceans began to slow and
our planet began to heal."

In a farewell letter to his former employees, Obama's outgoing energy
secretary, Steven Chu, wrote, "We don't want our children to ask, 'What
were our parents thinking? Didn't they care about us?'" After all, he
continued, "we do not inherit the Earth from our ancestors; we borrow it
from our children." As with so many of the folk quotations that gravitate
to the climate issue, this eternal wisdom, usually ascribed to the Amish
or Native American Chief Seattle, is actually nothing of the kind. It originates in a speech given in 1974 by the Australian Minister for the
Environment, Moss Cass, but don't expect any bumper sticker to give him
the credit.

These intergenerational challenges pull together several parallel cognitive themes. They seek to create proximity by showing how future events
follow from present choices and imagining the specific moment when
they might be brought to account. They avoid the problems of diffused
responsibility and bystander effect by creating a direct connection
between ourselves and those who will be affected. They build on our
hardwired sense of care for our children. And they bring in metaphors
from outside climate change, including wartime mobilization or roman-
ticized "tribal" lore.

They seem to be pressing all the right buttons. But do they actually
work?

The attitudinal research suggests that people who have children are no more concerned about climate change than anyone else—indeed possibly less so. A survey conducted in the United States, Canada, and Britain found that people with children were consistently less likely to believe that climate change was a serious threat, less likely to talk about it, and significantly less likely to have any opinion on how to deal with it. In the Canadian survey, people with children were 60 percent more likely to say that climate change was not really happening than people without children.

Another study in Britain concluded that having children has little or no influence and that the main determinant of people's attitudes is, as we would expect, their values and politics. If this disposes them to worry about climate change, then they are likely to loudly express a concern about their children's future, but this should not make us assume that this is a narrative that works across all boundaries.

And there are good reasons to suspect that having a child will mobilize the full tool kit of biases and avoidance strategies. Having children is usually an active choice in which we quite deliberately choose to highlight the reasons for having the child and suppress our knowledge about the world we might be bringing them into. Presuming that we wish the best for our children (and Sigmund Freud, who always liked to be controversial, claimed that elders subconsciously resented the young), this inclines us to an optimism bias concerning climate change and certainly concerning the prospects for our own children.

Professional climate scientists and campaigners (myself included) assume that our privilege will immunize our children from the worst impacts. For example, Peter Kelemen, a professor of earth and environmental sciences at Columbia University, expresses a relief that his own children are "lucky, among the 3 percent, talented, athletic, well educated." It will, he says, be the "people with less opportunity who find there is nothing they can do to help avoid destruction, displacement and despair."

We parents also prefer not to acknowledge that having children also involves making a huge contribution to climate change. A child in an industrialized economy will triple its parents' footprint; adding 9,441 metric tons of CO_2. It is on this basis that a childless friend—coincidentally the former campaigns director of a major environmental organization—justifies her annual flights to New Zealand.

So, the choice to have children compels us to write a narrative around climate change in which the overall prognosis becomes more optimistic, our own emissions become less significant, we become less vulnerable, and we accept a world of extreme inequality of future outcomes on their behalf.

And, of course, people with children can simply immerse themselves in the daily routine of tears, laughter, and the hunt for the missing shoe and put climate change into that category of tricky challenging things they would prefer not to talk about.

Even if we do not accept the moral responsibility, we are still preparing a set of alibis to defend ourselves against any future challenge: I did not know, I couldn't do anything, that was how things were, I did the best I could.

Or we will apologize. Institutional apologies have become a ritual in recent years. Bill Clinton apologized to the Hawaiians, African Americans, and Japanese Americans. The British have a culture based on insincere apology and, in recent years, Queen Elizabeth has said how sorry she is to the Sikhs and the Maoris—who still flashed their bottoms at her anyway. In return, the Fijians of Navatusila village have apologized most sincerely for killing a British missionary 140 years ago and eating "everything but his boots." Psychotherapist Ro Randall sees this fashion for political apology as directly linked to apocalyptic climate narratives. She says that both show a fear of vengeance, by people or by nature itself, and that in her view, apologies are directly related to a narrative of reparation emerging from guilt and loss.

However, it is also possible that future generations may not blame us at all—they may very well follow our lead and adopt exactly the same strategy of inactivity or indifference. For the moral philosopher Stephen Gardiner, it is this intergenerational element of climate change that makes it so morally dangerous. As he says, "each generation bears the full burden of previous inaction but will derive no benefit from their own action."

This is not to say that moral appeals based on intergenerational rights will not work, but they need to be carefully tailored to different values and cultures. In focus groups, Republicans strongly dislike moral demands to limit climate change but respond far more positively to messages about reducing our use of fossil fuels to provide a better life for our children and grandchildren.

But this does not mean that they respond well to attempts to put the two messages together. In fact, there are serious dangers that they could strongly resent the accusation that they are damaging their children's future. As the British climate denier and blogger James Delingpole says, "Kids can't sleep because they're so worried about the pets that are going to be drowned by the carbon monster." It's that drowning puppy bobbing to the surface again—which is not surprising, as the disastrous bedtime story commercial it references was clearly drawing on the "What did you do in the Great War, Daddy?" trope.

Speaking of which, despite its appeal as a piece of period design, the poster was not considered successful propaganda and was, by all accounts, loathed by the foot soldiers in the trenches—so much so that the artist Lumley later came to disown it. What really persuaded people to sign up was the campaign by Field Marshal Lord Kitchener to raise "Pals battalions" in which friends could join up, then die, alongside one another. Climate advocates would do well to remember that in the deadly social experiment of army recruitment, it was the combination of peer pressure, trusted communicators, social norms, and in-group loyalty that persuaded people to sign up—not a moralistic slogan, however clever it seemed to be.

36

The Power of One

How Climate Change Became Your Fault

WHAT MAKES CLIMATE CHANGE STAND out from *all* other global problems is that our individual contributions can be measured down to the last gram. We cannot identify our contribution to any other wicked problem, such as poverty, terrorism, or drug abuse—let alone quantify it. But with climate change, we can say with confidence whether our contribution is going up or down, how it compares with that of other people, and what changes would be needed to reduce it. Even though we do not make our cars, we still choose where to drive them. Even though we do not grow beef or asparagus, we still have the choice where and when to buy them, or whether to buy them at all.

The emissions that cause climate change result from decisions taken at *multiple* stages negotiated through the market. Ignoring wellhead production is a foolish error, but it is no less foolish to ignore the role of consumer decisions. People often feel powerless in the face of climate change, when, in fact, there is no other issue over which they have more personal control or involvement. Two-thirds of people in the United States and the United Kingdom and even more in Australia agree that individuals can and should actively reduce their personal contribution to climate change.

What is more, changing these personal behaviors may be the key to changing attitudes. According to "self-perception theory," behaviors are an important cue for self-image. So, if someone can be persuaded to adopt environmental behaviors, she may over time come to identify herself as someone with an environmental worldview. Green is as green does. And green does as green is.

On this basis, in the early 2000s, environmental organizations began to focus increasingly on the personal responsibility of consumers for climate change. It was a natural step for them, combining their long-held interests in consumer advocacy, personal responsibility, and ethical lifestyles.

To bring it home, they distilled personal actions into lists of household hints: some of them significant (reducing commuting, installing insulation and efficient heating), some of them marginal (not idling the car and turning appliances off standby), and some of them virtually irrelevant (not using plastic bags and unplugging cell phone chargers). The inclusion of such minor changes was supported by a large and verbose academic literature that, in its own jargon, argued that these "easy" steps could "spill over" into larger behaviors and that "hooking people with a small request" provides a "foot in the door," whereby they can be ushered onto the "virtuous escalator."

Books proliferated telling people to Measure their Carbon Footprint, get Low-Carbon, become Eco-Friendly, Save the Earth for a Fiver, Tread Lightly on the Earth, Kick the Fossil Fuel Habit, go on a Climate Diet. Or go on a Carbon Detox—the title of my own contribution to this short-lived and rapidly remaindered eco-tastic subgenre.

Maybe we all went too far and, in our eagerness to find homey messages that would engage people, we fell into the wicked trap of limiting climate change through the solutions we proposed. *An Inconvenient Truth* posited climate change as an existential threat yet petered out into a string of small options—changing lightbulbs, inflating tires, and driving a bit less. The Live Earth concerts in 2007 sought to fuel a global movement yet ended up promoting handy household tips. Six months after the concerts, I received a perky e-mail from the Live Earth team telling me how I could "re-use the heart-shaped candy boxes left over from Valentine's Day as picture frames, earring holders, or backpacks for dolls."

It was not long before governments started picking up on the theme. In the United States, lists of simple actions to prevent climate change were promoted by the Environmental Protection Agency, the Federal Highway Administration, state programs such as Cool California, and school curricula. The *New York Times* reported that kids, fired up on school eco-programs, had become "the little conscience sitting in the back seat," lecturing their parents about their behaviors and chanting, "Every day is Earth Day."

The unlikely leaders in this field were the thoroughly ungreen governments of Canada, Ireland, and Australia. In the early 2000s all three were intoxicated by an economic boom built on new roads, airports, and fossil fuel development. They duly ripped up their international commitments under the Kyoto Protocol or, in the case of Australia, refused to ratify it at all. And yet, strangely, all three countries then launched high-profile national campaigns to empower their citizens to take *personal* action against the global threat of climate change.

In Ireland, the Power of One campaign promoted the "breathtakingly simple idea" that each individual can "make a difference." In Australia, a twenty-million-dollar Climate Clever campaign targeted every household in the country. The Canadian government poured forty-five million dollars into national television ads for its One-Tonne Challenge, in which the comedian Rick Mercer harangued ordinary citizens to reduce their carbon dioxide emissions, shouting, "C'mon! We're Canadian! We're up for a challenge!"

On the face of it, these small steps appeared to be a means to avoid the political partisanship that bedevils climate change. Political action is messy and participatory. This seemed much more benign—speaking to values of national unity while enabling consumers to make well-informed decisions. We will fight it in the malls, we will fight it in the catalogs.

But that does not mean that when the governments focused on personal responsibility, they were not being political: They were being *extremely* political and framing climate change within a wider neo-liberal ideology that promotes private property rights and free markets. As the left-wing sociologist Ulrich Beck said, "We are all now moral entrepreneurs laden with personal responsibility but with no access to the actual decisions."

Worse still, these campaigns did not actually work. People were not up for the challenge at all and were certainly keeping well away from that virtuous escalator. Independent evaluations, now buried deep in the archives, found that the Power of One campaign was "only capturing those who were already converted" and the One-Tonne Challenge had achieved no change at all in overall energy use. In Australia, people became even less climate clever and a third fewer people considered climate change to be their most important issue after the campaign than they had before.

No one paid much attention to these brutal evaluations because these campaigns had never really been concerned with reducing emissions. In reality, they were a narrative gambit: to define climate change as a problem that lay at the very furthest end of the tailpipe in the purchasing decisions of the individual. Behind their uplifting slogans, and their appeal to national unity, what they were really saying was "climate change is *your* fault."

And here lies the problem. As soon as one creates responsibility, one creates blame. Blame creates resentment, and talk of responsibility in the home makes that resentment very personal indeed. What none of us fully appreciated at the time was how readily these anodyne messages would be mobilized to fuel people's sectarian prejudices.

Conservatives in particular loathed being told what to do by governments and liberal environmentalists. In one revealing experiment by a team at the University of Pennsylvania, many conservatives refused to buy a low-energy lightbulb once the packaging carried a sticker reading "protect the environment." At my Texas Tea Party meeting, Craig recalled how he challenged an environmentalist by saying, "You're on this computer and you're using electricity made from the coal that you claim you hate, dug out of the ground by the man who you are telling he isn't allowed to eat meat." He received the loudest cheer of the night.

There is a deep irony in this. Research by the psychologist Jonathan Haidt into moral foundations of different worldviews found that it is conservatives who have the greatest moral emphasis on personal responsibility and that it is liberal environmentalists, with their highly individualized values, who are actually the group least suited to working together for a shared goal.

It was not just conservatives—left-wing trade unionists were just as repelled. Running focus groups with activists from one of Britain's largest trade unions, I found that few things irritated them more than the phrase "lifestyle change," which for them was poisoned by an association with middle-class environmentalism and government buck-passing. There was something very wrong here: Surely, I thought, trade unionists, of all people, would respond to a call for collective action against a common threat?

Somehow, this was all the wrong way around. Those campaigns

urging people to take personal responsibility and work together to "save the planet" were saying the wrong things to the right audience and the right things to the wrong audience. So much for the hope that small personal lifestyle changes might shift people's attitudes and bring people together; if anything, they seem to reinforce people's prejudices and drive them apart.

This is because our willingness to make a personal sacrifice is entirely bound up with our sense of social identity. If we feel an affinity with the group, then we will willingly make a contribution to prove our loyalty. In times of conflict, we may even sacrifice our life. But this strengthened sense of in-group identity, and our socially wired sense of fairness, makes us deeply resentful of moral rules laid down by outsiders that they themselves do not appear to follow.

Nor do small changes in lifestyle necessarily lead onto the virtuous escalator to larger commitments. Further research has found that even those people who accepted the threat of climate change were all too ready to adopt a single simple action as a token of their concern and then go no further. Columbia University psychologist Elke Weber has identified numerous examples from farming, health, and politics in which people respond to a problem with what she calls "single action bias." She argues that this may be another bias derived from our evolutionary past when threats were simpler and a single short-term action could safely relieve us from danger and the anxiety of worry.

People then use that single act as a personal justification—what psychologists call moral license—to offset further damaging behavior, just as people order a supersize Diet Coke as an antidote to their double bacon cheeseburger. So, research has found repeatedly, people who buy energy-efficient lights and appliances tend to use them more. People who insulate their houses then turn up the thermostat.

They transfer the moral license to other areas too. When residents in a Boston apartment building were sent notes (with pretty leaf graphics) asking them to save water to "help preserve the environment," they used 7 percent less water. And they then used 6 percent more electricity.

Researchers at the University of Toronto found that this moral licensing effect was so powerful that people who had just bought environmentally friendly products in an experiment became markedly more willing to take up an opportunity to cheat the university and even steal money.

Within the issue of climate change, people use moral licensing as part of a deliberate process to write narratives that diminish their own responsibility. In interviews, people exaggerate their own small actions and portray them in heroic terms. One participant in a British focus group boasted that he recycled everything he could and that not one piece of paper went into his garbage. This, he added, "makes me feel less guilty about flying as much as I do."

So, once again, climate change has become wickedly defined by its solutions. For people who accept that climate change just might be a major threat, providing solutions based on small lifestyle changes makes it seem far less dangerous and carbon comes to seem like another form of litter that they really shouldn't drop.

For people who doubt that climate change exists, demands to change their lifestyle confirm their suspicion that the real threat comes from the environmental liberals who want to control their lives.

What is missing, and what is urgently required, is a coherent policy framework that provides a contract for shared participation—whether through voluntary measures or, as many campaigners now demand, some form of tax, ration, or dividend—within which personal actions are recognized and rewarded alongside equally important contributions from government, business, and fossil fuel companies. Not the power of one, but the power of all.

37

Degrees of Separation

How Climate Experts Cope with What They Know

Sitting on a panel at the 2012 annual assembly of the Tyndall Centre for Climate Change Research, I remember vividly the palpable unease that entered the room when the discussion turned to four degrees. There was a hesitation in broaching the subject. Voices became quieter and less confident. Nonetheless, everyone in the room talked about four degrees Celsius of warming as being entirely probable if not unavoidable.

At the reception later that evening, the scientists chatted amiably in pairs and small groups, clutching their glasses of warm white wine and balancing plates of canapés. With their slightly rumpled old-fashioned clothes and their polite, intense demeanor, they looked like any other group of highly educated professionals—to my eyes, rather like the audience for a concert of somewhat challenging chamber music.

However, listening to the detail of their conversations reminded me that this gathering was far from ordinary. The people in this room constituted a large part—maybe even the majority—of those in Britain who truly understood why a global temperature increase of two degrees might, just, be manageable, and why one of four degrees would be an utter catastrophe. This, after all, is what they have spent their lives studying. Of all people, they know all too well that the phrase *four degrees* is shorthand for environmental, social, and economic collapse. And, as their models keep telling them, we are heading straight for it and could well reach it within sixty years.*

According to Professor Lonnie Thompson, a climatologist at Ohio State University, those in his profession are a stolid group, not given to "theatrical rantings about falling skies." However, he says, they now feel compelled to speak out about the dangers because "virtually all of us are now convinced that global warming poses a clear and present danger to civilization." Extraordinary though this statement is, even more extraordinary is that it appeared in an otherwise sober report in the journal of a respected international science association.

Every year their warnings have become ever clearer and more serious. And, it seems, every year they have become less believed. Returning to the quote with which I started this book, these scientists are, I fear, uncomfortably similar to the handful of people in 1942 who knew what was happening to the Jews of Europe, who carried the weight of that dreadful knowledge but struggled to persuade anyone of the existence of a crime of such immensity.

One scientist told me that he was so disturbed by the latest findings that he wrote to a few close friends—he named some of the world's most senior scientists—and asked them: the future of humanity depends on this, is there any chance—please any chance—that we could be wrong? They replied immediately, saying that they too constantly worried about this and (contrary to what the skeptics claim) were always open to the possibility of being wrong. However, whenever they went back over the evidence, they could not avoid the uncomfortable conclusion that they had indeed gotten this right. "We are active fatalists," he told me.

Activists and campaigners also struggle with this sense of anxiety, suffering sleepless nights and panic attacks. Dorian Williams, an anthropology senior at Brandeis University who leads the campus divestment campaign for 350.org, says that she experiences "very serious, very low states of being for hours to days to weeks at a time." It's never going to go away, she says, but "you just have to work through it so that you can keep fighting."

People who deal every day with climate change as a reality provide an

---

* I discuss some of the implications of a four-degree temperature increase in the final chapter of this book.

important insight into the ways that humanity as a whole will cope with its psychological and moral challenges. Almost all analysis concerns the psychology of people who refuse to accept the science—which, understandably, they resent. But what about the people who are already convinced? They are the advance party and, as climate impacts build, everyone will follow their lead.

Their internal moral dilemmas come to a head as they struggle to square what they know about the impacts of high-carbon lifestyles with the pressure to conform to a society where those lifestyles are not just encouraged but also often required as a mark of social belonging.

I have an informal social research project—life is one long experiment after all—in which I gently coax climate change experts to talk to me about their personal holidays. A senior climate economist at the World Bank admitted that he flew regularly for breaks in South Africa but said that that this was a force for good because the carbon offsets he bought "help set a price in the carbon market." A national media environment correspondent decided to fly with his family to Sri Lanka because, he said, "I can't see much hope." A climate scientist specializing in polar research takes several long haul flights every year for skiing holidays because the "job is so stressful." The lead climate campaigner for one of the largest U.S. environmental organizations flew so often for her work that she could take regular long-distance holidays using her air miles with an automatic upgrade to business class.

All of them felt uncomfortable discussing their leisure flying, and I have found that there is a norm of silence—a meta-silence even—around this topic. Nonetheless, when prompted, all of them could present complex narratives to justify their own behaviors, often containing a moral license or deferring to the social norm among their fellow middle-class professionals. They all argued that they would gladly stop flying but—and here they drew on their insider understanding of the scale of the problem—a single personal sacrifice is meaningless unless it is supported by wider systemic and social change. Ironically, their own well-informed arguments provide the clearest evidence possible that scientific information, on its own, is unable to counter socially engrained behaviors.

Professor Kevin Anderson, the former director of the Tyndall Centre, is unusual for his reluctance to fly for any reason. His audience at a recent conference in China was astonished and impressed when he told them he had come (and would return) by train. He is convinced that this added to the legitimacy of his science.

Anderson regards it as "incredibly disturbing" that the people who shape climate policy are such profligate fliers. He tells me of a conversation with the director of one of the largest power utilities in Britain, who told him, quite casually, that the following weekend he was flying with his horse to China to go riding. Anderson explodes, "We were both about to give evidence at a government hearing on climate change and he was flying his bloody *horse* to China! . . . And when I challenged him, he looked at me like I was some kind of radical lefty!"

Experts seem to believe, Anderson tells me, that the pearls of wisdom that they've rained down from thirty-two thousand feet in a first-class seat are so important that they outweigh their emissions and those of the people like them. They don't see that the reason we have this problem is precisely because of people like them and, he adds, being more conciliatory, "people like me."

And, I should also add, people like *me*—because I am an expert flyer too. I fly rarely and I always try to justify each flight. But as that word *justify* reveals, I am also prone to constructing a narrative that can resolve the inner conflict I feel every time I sit on a plane. It is all immensely frustrating because I must admit that I love travel and, in my pre–climate change days, I flew a lot. So I know very well that flying is addictive.

Mark Ellingham, the founder of the Rough Guides travel books, coined the phrase "binge flying," which he compares with nicotine addiction. Interviews with frequent travelers find them using the same language as other forms of addiction. They talk about the buzz or rush, their loss of inhibitions, finding new meaning in life, and their depression on their return.

Maybe this is why the self-serving narratives we experts mobilize to justify our personal flying are so uncannily similar to those that people raise around addictions: I need to do this, I'm not hurting anyone, everyone else does it, I've worked for it, I can stop anytime, other people are far worse.

Texas state climatologist John Nielsen-Gammon says that the public needs to remember that the people who work on climate issues may be smart but are still human beings like any others, "driven by varying mixtures of ambition, curiosity, orneriness, self-confidence, and altruism."

However, climate experts *are* different from other people in one critical aspect: We are the lead communicators of climate change and our own actions will always be monitored as a measure of our trustworthiness. In other areas, inconsistent behavior by decision makers is utterly relevant: the racial prejudice of judges, the tax evasion by politicians, and the sexual behavior of priests are all matters of intense public attention because we know intuitively that an internal conflict may undermine their judgment.

Inevitably we run the risk that we will project our own values, inconsistencies, and silences onto the story we tell. Is it any surprise, given these internal conflicts, that there is so little mention of flying among the list of personal actions promoted by environmental groups and the U.S. Environmental Protection Agency? Or, indeed, that international aviation is not included in national emissions calculations or the Kyoto Protocol?

Renee Lertzman, a visiting psychology fellow at Portland State University, argues that it is mistaken to judge these inconsistencies as arrogant or hypocritical or apathetic. They are, she says, best understood as a strategy by which experts defend themselves against their anxiety and the internal dilemmas that cause them pain. "We cannot tolerate our own complicity, so we externalize and project our concern onto others—the airline industry or the failure of government policy to control it," she tells me.

She recalls a participant in one of her workshops complaining that people who fly "lie" when they say that they care about climate change. No, she stresses, "it is *not* lying—these are intentions that they are struggling to negotiate." Nor, she says, is there a gap between what they say and what they do. She prefers to see this as a tangle of conflicting needs, or, she suggests, a tapestry.

Rosemary Randall, a psychotherapist who has worked extensively with climate scientists, says that she frequently encounters their "bewilderment, depression, and despair at public attacks or indifference." Their solution, she suggests, has been "to move further into the world of reason—more graphs, tighter arguments, greater precision."

Another psychologist, who works alongside climate scientists in one of the largest British research councils (and so preferred to talk to me off the record) is perpetually disturbed that her colleagues fly constantly and never talk about their anxiety or the implications of their work. She is convinced that, as a result, generating ever more knowledge has becomes the end in itself. They have, she told me, created "a huge information machine run by experts, reinforced by other experts, and all they do is sit around in expert committees, and make their expert presentations to each other."

This rationalist expert culture protects scientists from the emotional content of their work. When Lertzman interviewed scientists at the Environmental Protection Agency in 1998, she expected them to share stories about their emotional struggle at the frontline. She tells me she was really surprised to hear them say, "'I am a scientist and I don't engage on that level.'"

In the fascinating article "When Swordfish Conservation Biologists Eat Swordfish," the marine biologist Giovanni Bearzi complains that biologists who spend their professional lives researching unsustainable fishing can nonetheless sit down at a restaurant and order swordfish or tuna from those declining stocks. It is, he says, "as if monks advocating poverty were to wear jewelry and expensive silk robes."

Yet, if Lertzman and Randall are right, we could see this in a quite different light. When people gratuitously perform the thing they warn against, it suggests a ritual of public disavowal. They are managing their own emotional anxiety by policing a strict cognitive divide between work and play, information and responsibility, the rational brain and the emotional brain. Activists often quote the motto "be the change you wish to see," which they ascribe to Mahatma Gandhi (although, of course, he never actually said this). In a way, these experts are also acting out the world they wish to see—a world in which they do their job, governments do their job, resources are managed sustainably, and then they can fly to Italy on holiday and have that well-earned swordfish steak—goddamnit!

Professor Chris Rapley, former director of the Science Museum in London and one of Britain's most senior climate scientists, has become an unlikely advocate for the psychoanalytic arguments, which he gladly defends against the positivist prejudice within the science community that "psychotherapy is not rigorous and quantitative."

Rapley speaks with remarkable honesty and clarity about the internal stress he endures from what he knows. "It is," he tells me, "*so* difficult to be optimistic, however much you argue yourself into an optimistic position. I know I have tended to deal with my own anxiety by placing what I know into watertight compartments. The fact that we climate scientists can sleep comfortably at night tells you that we have unconsciously worked very hard at this."

Lertzman and her fellow psychotherapists argue that we are all irrational, unconscious, confused human beings and we are *all* struggling to
make sense of this issue. This is why she finds the cognitive explanations
for our avoidance of climate change to be "incredibly limited." They put
the blame on the "ignorant, self-centered, shortsighted people, in contrast
to the enlightened and evolved people." The focus on political affiliation
is also superficial because it does not explore what leads people to become
so strongly identified with those affiliations. Hatred, she says, is always a
clue that something else is going on.

For Lertzman, the argument that climate change is too hard for us to
deal with is "ridiculous," and if we turn it on its head, "there is plenty of
evidence that we have enormous capacity for deep care and concern." The
question is then how to reframe the argument away from the gap and
into the tangle of the tapestry. People need to be in the place where their
anxieties are recognized, to be able to say, "Yes, this is scary; this is hard,"
and only then, she says, can we be truly mature, creative, strategic, and
innovative.

38

Intimations of Mortality

Why the Future Goes Dark

THE JACOB K. JAVITS CONVENTION Center in New York City is heaving—115,000 fans crammed in for the second day of Comic-Con, the largest comics convention on the East Coast. I am here to ask a simple question: What do you think the future will be like?

My reasoning is this: These people are young, smart, and curious about technology and future worlds. Surely, as they stand around in lines waiting for autographs, they will have spare time to answer a few questions from a stray British social researcher—not in costume, although one woman eyes my scruffy trench coat and asks, "Out of curiosity, have you come as Inspector Gadget?"

So, I ask them, what *do* you think the future will be like?

The surprise is that they have little idea or, it would seem, desire to find out. One woman says, "I've never thought that far ahead—I like living in the present." A man farther down the line is concerned that it might be "a one-color-jumpsuit kind of future." "Like *Logan's Run*," he adds when I look perplexed.

Brian Ferrara is selling nine-hundred-dollar replica weapons from science fiction video games. "I'm not a doomsday prophecy kind of guy, but I am a realist," he says. So, being realistic, he doesn't see a bright future, but he is very vague about the details. Maybe, he speculates, we will be immobilized, strapped to a chair with a feeding tube.

One couple are more politically alert, having spent time with the

205

Occupy movement. They anticipate some kind of corporate dystopia, But, they say, there are other issues too. Overbreeding. The constant battle over fertility rights. "Yes," says the woman, warming to the theme. "Politicians! Get out of my uterus! Leave my lady parts *alone!*" In her one-piece latex Catwoman outfit, she looks reasonably safe for the moment.

And climate change? In over twenty interviews, not one person mentions climate change until I prompt them to do so. Then they have lots of views. No one doubts that it is happening or is going to be a disaster. "It will escalate into catastrophe." "If we can't cope with that, we'll all die like the dinosaurs." But asked to identify when these impacts might hit, they reckon it's still a long way off. "Maybe my great-grandchildren will have to deal with it," Catwoman says.

Whatever happened to the future? Even when I was growing up, a final straggler in the baby boom party, there was no doubt about what the future would look like. Shiny glass buildings, food pills, mining the oceans, monorails. Images of the future were everywhere. Now, research shows, people are unwilling to even think about it.

Bruce Tonn at the University of Tennessee, Knoxville, has spent the past decade asking people what they think will happen in the future. According to his research, most people interpret "the future" as being no more than fifteen years away and, beyond twenty years, people's ability to imagine the future "goes dark." It's an intriguing term, and I ask Tonn to explain it. He says that people are "just not able to imagine *any* type of future. They cannot visualize their lives, cannot visualize society, cannot visualize the impacts of various policies or lack of them." Tonn's research revealed a deep, underlying pessimism. Almost half of his respondents would not wish to have been born in the future and anticipate that humanity will go extinct, most likely from environmental collapse.

A survey of five hundred American preteens found that more than half felt the world was in decline and a third believed it would not exist when they grew up. In Australia a quarter of the children believe that the world will come to an end before they reach adulthood.

Extinction is an emerging narrative around climate change—not just extinction generally, but our own extinction specifically. *A World Without Us*, Alan Weisman's book on the environmental recovery of a world emptied of humans, became a bestseller. In books such as Fred Guterl's *The Fate of the Species* and Clive Hamilton's *Requiem for a Species*, climate change writers argue that our extinction is the ultimate danger.

Elizabeth Kolbert's book *Field Notes from a Catastrophe* ends with the line, "It may seem impossible to imagine that a technologically advanced society could choose, in essence, to destroy itself, but that is what we are now in the process of doing."

Recent years have also seen the emergence of a new field of extinction studies. The Future of Humanity Institute, nested within the Faculty of Philosophy at Oxford University, specializes in the study of catastrophic risks that threaten the future of humanity. The rival Cambridge University has founded the Center for the Study of Existential Risk with the support of Jaan Tallinn, multimillionaire cofounder of Skype. This is, one realizes, cool stuff.

The Future of Humanity Institute conducted a poll of academic experts on global risks. They gave an estimate of 19 percent probability that the human species will go extinct before the end of this century. *The Stern Review: The Economics of Climate Change* factored a 9.5 percent risk of extinction within the next century into its calculations.

Extinction fits neatly into an altogether more flippant and fatalistic narrative that it is too late to do anything. As the late comedian George Carlin put it:

Save the planet! What!? Are these fucking people kidding me!? The planet isn't going anywhere. We are! We are going away, so pack your shit, folks. We wouldn't leave much of a trace either. Just another failed mutation, just another closed end biological mistake. The planet will shake us off like a bad case of fleas. A surface nuisance.

Or, in a calmer version of what is essentially the same narrative, Zen master Thich Nhat Hanh tells us that "the collective anger and violence" of global warming will lead to our destruction within a hundred years, but we can accept this because "Mother Earth knows she has the power to heal herself."

These narratives do not belong exclusively to any single identifiable worldview. They are experimental, building momentum as people try them out and pass them around. They are already widespread. Carlin's nihilistic monologue has been viewed more than five million times on YouTube.

If this is a defense mechanism, it is one that bypasses the entire issue of our moral responsibility. It is as though we have moved from a bedside vigil to bereavement counseling without actually experiencing the death itself.

And when it is said that we are the ones who will go extinct, this is the most slippery of all *we*'s. As I discussed earlier, people are consistently far more optimistic about their own chances than they are about those of humanity as a whole. No one using this language seems to seriously consider that they or their own associates are under direct threat—these frames prepare them to accept the suffering of others as unavoidable and required.

Climate change imagery draws heavily on the iconography of death. Starvation, cracked earth, skulls, dead trees. Reports of extreme weather events give prominent mention to the numbers of deaths, highlighted in headlines about "killer heat." Metaphors of terminal illness, cancer, and murder appear regularly in environmental articles. An article in the *Sydney Morning Herald* framed climate change as a terminal illness under the headline "Prognosis for a Planet: Death."

The veteran environmental scientist James Lovelock has explored these metaphors in depth. A climate scientist, he says, is like "a young policewoman" who has "to tell a family whose child had strayed that he had been found dead, murdered in a nearby wood." Science study centers are "the equivalent of the pathology lab of a hospital" and are reporting that Earth "is soon to pass into a morbid fever and soon her condition will worsen to a state like a coma."

Beyond such attention-grabbing associations between climate change and death lies a more interesting question: Does climate change *in its essence* trigger our fears of our own death and is our response shaped by those fears?

The anthropologist Ernest Becker argued that a fear of death lies at the center of all human belief. The denial of death, he argued, is a "vital lie" that leads us to invest our efforts into our cultures and social groups to obtain a sense of permanence and survival beyond our death. Thus, he argued, when we receive reminders of our death—what he calls death salience—we respond by defending those values and cultures.

Becker's theory, called terror management theory, has been supported by more than three hundred experiments and predicts that when people are made directly aware of their death, they immediately rationalize the threat, often by denying the personal risk or the proximity, just as smokers will say, "It's still a long way off, so I may as well live a little."

Janis Dickinson, a professor of neuroscience at Cornell University, places climate change within the thinking of Becker. She suggests that many of the standard responses to climate change, of extreme rationalization, denial, or placing climate change impacts far in the future, are all consistent with our responses to our fear of death.

We cannot stand to think of the death of our own children, but we accept that they will die after we ourselves have died. Similarly, we can avoid the fear of climate change by placing its impacts beyond our own life span. In focus groups, people often do this quite openly, justifying their indifference with the observation that it is all in the future, when they will be long since dead and gone.

However, there is another, subtler aspect to Becker's terror management theory. When the reminder of mortality is subtle or so subliminal that people do not even notice it, they display a greatly enhanced sense of the superiority of their own social group, and that can lead them to give increased attention to status, money, and improved self-image. Becker believed that our innate way of coping with our death is to invest our energy in our social group and its achievement—what he called our "immortality project."

Dickinson suggests that it is the subconscious associations of climate change with death that are having the greatest effect, fueling the extreme polarization of deniers and believers, and driving the wider population toward status-driven high-carbon lifestyles.

She cites the strong evidence that people interpret images of environmental destruction in terms of their own death. As Becker's theory predicts, exposure to images of death increases environmental concern among those who already have those values, and reduces it among those who do not.

Throughout this book, I have avoided arguments that are not supported by strong evidence, but in this case some conjecture is justified, not least because so many thoughtful people (Bill McKibben, Bob Inglis, Daniel Kahneman, and Joe Romm, among them) have spontaneously suggested in my interviews that climate change might be a proxy for death.

The environmental author Carolyn Baker finds it impossible to avoid this connection: "Collapse forces us to march in a funeral procession toward the end of life as we have known it—and the end of ourselves as we have known them. And who, I ask, would willingly sign up for this?"

Many people who work on climate change deal with a deep sense of grief. Journalist Ross Gelbspan wonders why he has such problems crying when he feels such a terrible sadness to see the loss of the future for young people who look forward to fulfilling their lives. "Instead of just running away from it, I try to take a deep breath and close my eyes and let it in."

The campaigner Bill McKibben agrees that climate change does feels uncannily like our own death. When I invite him to explore the theme, he adds an important caveat: This is, he tells me, quite unlike a natural death. "We are grieving for what we are doing and our own inability to deal with it. We all know we are going to die, and we used to be able to cope with the thought that our life was contributing to something larger that would survive us. Now even that has been taken away from us." So even the "immortality project" that compensates for our own deaths has been taken away from us.

Increasingly we are told that whatever we do, we are *committed* to some uncertain future catastrophe that threatens to render the past meaningless. All we can do is wait for it to come. It feels both real and unreal, something we are told will happen, that we might rationalize but we can't quite believe.

This strange sense of impermanence was the central theme in McKibben's seminal book on climate change, *The End of Nature*: "Our comforting sense of the permanence of our natural world, our confidence that it will change gradually and imperceptibly if at all, is, then, the result of a subtly warped perspective . . . We are at the end of nature."

Maybe it is appropriate to leave the last word to the founder of psychotherapy, Sigmund Freud, whose work so often revolved around the centrality of death in our psyche. In his short essay "On Transience," Freud explores the way that our anticipation of future death diminishes our view of the present. In 1915 Freud was walking with a friend in the summer woods, a few months after one hundred thousand men had been massacred in the Battle of Ypres:

> The poet admired the beauty of the nature around us, but it did not delight him. He was disturbed by the idea that all this beauty was bound to fade, that it would vanish through the winter, like all human beauty and everything beautiful and noble. All the things he would otherwise have loved and admired seemed to him to be devalued by the fate of transience for which they were destined.

39

From the Head to the Heart

The Phony Division Between Science and Religion

TIM NICHOLSON'S MODEST AND SOFT-SPOKEN style provides a disconcerting cover for an altogether more flamboyant and risk-taking personality. In 1995 Nicholson and ex-army wife Major Jo became local celebrities after they drove from their hometown of Oxford, UK, to Oxford, New Zealand, in a 1954 Morris Oxford—a bulbous British car that looks like a boiled sweet and is about as powerful as a lawnmower.

In 2009, Nicholson was in the news again—this time around the world—when he sued his former employer, a large housing organization, on the grounds that it had fired him from his position because of his deeply held conviction in climate change. Nicholson built his court case on European legislation that protected workers against discrimination on the grounds of "any religious or philosophical belief." He had, quite deliberately, begun another dangerous journey: this time right through the minefield that lies between those who regard anthropogenic climate change as an irrefutable scientific fact and those who see it as an ideologically driven belief.

Skeptics saw the case as confirmation of their long-held argument that climate change was a new and false religion. Environmental campaigners applauded his bravery, and one newspaper, unhelpfully reinforcing the religious theme, declared him to be a green martyr.

Scientists were a lot less convinced. Science writer Wendy Grossman said he should be "appalled" by the case he had brought. She wrote, "Science is not a belief system but the best process we have for establishing the truth. If the issue of climate change is one of competing religious beliefs, then those claiming impending doom can be safely ignored."

Nicholson would never argue that climate change itself was similar to a religious belief. What he was arguing, and what ultimately won him the case, was that this scientific evidence could become the basis of a life-changing moral philosophy and that this was *similar* to many religions—based on principles of caring for others, responsibility, and thoughtfulness. As Nicholson told me, "In the end, climate change is not some facts and figures; it comes down to what's in your head. And that's a belief."

Most climate scientists hate to talk about *belief*, which they regard as diametrically opposed to reality-based facts. Adam Frank, a professor of astrophysics at the University of Rochester, says, "I always feel a bit weird when someone asks me if I 'believe' in climate change, as if it's the Easter Bunny or Santa Claus." Australia's chief scientist Ian Chubb complains, "I am asked every day 'do you believe?' and every now and then I make a mistake and say yes or no. But it's *not* a belief. It's an understanding and interpretation of the evidence."

As with so many of the arguments that surround climate change, this is not really about the word *belief*, but about the religious frames that it triggers and the false polarity it suggests between the rational brain and the emotional brain. In the struggles with deniers, the word *belief* has become poisoned, and many scientists see it as the antithesis of peer-reviewed science. This is why I prefer to use the word *conviction*—to indicate a condition of strongly held opinion, reached through a personal evaluation of the evidence.

Many deniers harbor a deep hatred for all religion, seeking to smear climate change by association. To one business columnist, climate advocates are like "crazed American televangelists who predict that the Antichrist will come next Tuesday or that God will purge the land of homosexuals."

The metaphor is especially strong in Australia. Ian Plimer, a retired petroleum geologist who has built a lucrative new career as a leading Australian denier, based an entire book, *Heaven and Earth*, around this theme, arguing that climate change "creates a fear of damnation, demands appeasement by selling indulgences to the faithful and demonizes dissenters." Even Cardinal Pell, Australia's most senior Catholic, describes emissions reductions as "religious sacrifices" and compares the sale of carbon credits with "the pre-Reformation practice of selling indulgences."

However, climate skepticism is, in a manner of speaking, a broad church, and it also includes those, especially among the American Christian right, who see climate change as a heresy that "speaks to the inherent spiritual yearnings of human souls and seduces children in our classrooms through spiritual deception." These are the words of Calvin Beisner, the founder of the Cornwall Alliance, which markets a set of twelve DVDs that will "provide the armor" to rise up and slay environmentalism, or as he calls it, the Green Dragon.

In 2006 Beisner and twenty-two evangelical leaders launched "An Evangelical Declaration on Global Warming," arguing that it is a natural cycle. One of the most active promoters of the declaration, Bryan Fischer of the American Family Association, argues that we have a God-given right, indeed requirement, to burn fossil fuels because "the parable of the talents tells us that the wicked and lazy steward was the one who buried his talent in the ground and did not do anything to multiply it."

However, conservative evangelicals, like the political right as a whole, are split between those who think that climate change (if happening at all) is due to natural cycles and those who accept that it is due to human behavior and that taking action to prevent it is the moral equivalent to protecting unborn life and preserving the family. So says the Evangelical Environmental Network in its rival manifesto, "Climate Change: An Evangelical Call for Action." The Network achieved widespread attention for an inspired television commercial about the environmental impacts of car travel that ends with the question "What would Jesus drive?" There are now similar initiatives among Jews, Muslims, Catholics, Buddhists, and Hindus that weave climate change into their own narratives and traditions.

In spite of this, what has been remarkable is how *little* involvement religions have had in the climate change issue. Previous social justice movements, from the anti-slavery campaigns through civil rights, anti-apartheid, anti-debt, and anti-poverty campaigns, arose through church networks.

People of faith have found it hard to incorporate this new issue into their existing worldview. Climate change is seen as an environmental issue that is poorly defined and contested in their theology. For

214 DON'T EVEN THINK ABOUT IT

conservative Christians, it is tainted by its association with the liberal environmental movement and has become bundled among the checklist of issues that define their group loyalty.

Environmentalists are equally wary of religion and seem to form strategic alliances with just about anyone before they talk to religious groups. This is a major tactical error. All of the world's major religions are growing, Christianity and Islam fastest of all, and much of that growth is from the more fundamental strains of their faiths. Within the United States, only 5 percent of people are members of environmental organizations, but more than 70 percent of Americans still identify with a religious faith, and more than a quarter of Americans consider themselves to be born-again or evangelical Christians.

Even Christians who do care tend to keep their faith and climate change "in two separate boxes," says Erin Lothes Biviano, a professor of theology at the College of Saint Elizabeth in New Jersey who spent a year interviewing climate change campaigners in the faith communities. She tells me that they rejected the comparison between climate change and religion because "they have an experiential relationship with their faith that is special, and they would not say that climate change has that same personal luminous quality."

So, what association between religion and climate change *is* appropriate? In one sense, they are clearly incompatible. Religions are based in ancient texts and revealed knowledge. Climate change is grounded in constantly changing and carefully evaluated scientific data. Religions relate to the otherworldly, the spiritual, and the afterlife. Climate change is utterly worldly in its causes and solutions and offers nothing spiritual.

However, climate scientists with strong religious faith argue that this has always been a false divide. Katharine Hayhoe is the director of the Climate Science Center at Texas Tech University and is also an evangelical Christian who is married to a pastor—an unusual combination that has led *Time* magazine to list her as one of the World's 100 most influential people. Hayhoe says, "The facts are not enough. When we look at the planet, when we look at creation, whatever it's telling us is an expression of what God has defined it to be. So instead of studying science, I feel like I'm studying what God was thinking when he set up our planet."

Sir John Houghton, who founded and then chaired the Intergovernmental Panel on Climate Change for fourteen years, is also a preacher in the Methodist Church. In 2002 Houghton hosted a conference between scientists and U.S. evangelical leaders at Oxford University (because he was told that "Americans love coming to Oxford"), which was the first attempt for a conservative audience to talk with one another about climate change using the language of faith. The conference was a resounding success and initiated a change of heart in participants that many later described as a conversion. The initiative provides further strong evidence that even the most unconvinced people can be persuaded by trusted peers who understand their values and can use their common language.

Like Hayhoe, he says that his religious belief and his scientific research are entirely compatible. God, Houghton tells me, creates the laws, and his role as a scientist is to discover them. He recognizes that scientists talk about the evidence base rather than the belief, but then, for Houghton, religion is also evidence-based. "Even if there are aspects that you do not understand, it all fits together in a way that you cannot escape from and there are laws of evidence to support it."

People of religious faith have understood all along that there is actually no clear dividing line between the rational and the emotional brains, but rather a conversation between the two. As the Ecumenical Patriarch of Constantinople Bartholomew said, "We know what needs to be done [about climate change] and we know how it must be done. Yet, despite the information at our disposal, unfortunately very little is done. It is a long journey from the head to the heart; and it is an even longer journey from the heart to the hands." This is another expression of the challenge of converting the rational-brain understanding of climate change into the emotional-brain commitment to action.

For the purposes of this book, though, what makes religious belief so relevant to climate change conviction is that both struggle against the same cognitive obstacles. As I have already discussed, climate change is extremely challenging because it requires people to accept that something is true solely because of the authority of the communicator, it manifests in events that are distant in time and place, and it challenges our normal experience and our assumptions about the world. Above all, climate change requires people to endure certain short-term losses in order to avoid uncertain long-term costs.

Religion faces every one of these obstacles, but to an even greater degree. It is even less certain, has none of the objective proof of science, is based on evidence that is remote from people's ordinary existence, and requires people to accept rules governing their most intimate lives—their sexual activities, diet, and child rearing. It has, I grant you, the major advantage of offering personal reward in an afterlife, though this too is based on extreme uncertainty.

As the Reverend Sally Bingham, an Episcopalian preacher and renewables advocate, put it to me: "We believe that Mary was a virgin, that Jesus rose from the dead, that we might go to heaven. So why is it that two thousand years later, we still believe this story? And how can we believe that and not believe what the world's most famous climate scientists tell us?"

Religions also call on people to constrain their wordly desires. The tradition of abstinence and self-restraint works through all the world's great religions: Eastern and Western. To quote Muhammad: "What have I to do with worldly things? My connection with the world is like that of a traveler resting for a while underneath the shade of a tree and then moving on."

Religions embody long-term thinking, encouraging their members to accept responsibilities and invest in a legacy that extends far beyond their own lifetime on Earth. The tagline of the Coalition on the Environment and Jewish Life, for example, is "protecting creation, generation to generation."

Above all, religions have found ways to build strong belief in some extremely uncertain and unsubstantiated claims through the power of social proof and communicator trust. Few are less certain, or more successful, than Mormonism, which has becomes the fastest-growing religion in the United States.

Mitt Romney, former governor of Massachusetts, was the first Mormon—a ward bishop, no less—to run for the presidency. He was also the first candidate to openly repudiate climate science. Which raises a very interesting question: What are the key differences that can lead a highly intelligent and worldly man to say "I am uncertain how much of global warming is attributable to man" and yet accept as *certain* that a transcription of tablets found buried in a hillside contains the word of God. I am not seeking to mock Mormons, just asking a legitimate question: What is it that makes one irrelevant and fraudulent and the other the rock of a man's life?

Maybe the question, then, is not whether climate change is too much like a religion, but whether, in our determination to keep the two apart, we have ignored the most effective, tried, and tested models for overcoming disbelief and denial.

40

Climate Conviction

What the Green Team Can Learn from the God Squad

THE IMAGES ON THE VIDEO screens start in familiar documentary style with some low bass tones, a plinking piano, a sun rising, and a slow-motion hand running through the sand. It feels strangely reminiscent of the opening to *2001: A Space Odyssey*—not, I imagine, a favorite film of the twenty-five thousand evangelical Christians who are now rising to their feet, clapping and cheering, as the bass riff picks up and the thirteen-piece rock band rises through the floor of the stage. "Do you hear that beat? Do your *hear* that *beat*? That's the beat of the FREEDOM!" "YAAAAAAY," we all go.

Lakewood Church, the largest church in the United States, offers a great package. Great venue. Great tunes. Great gift shop. Pastor Joel Osteen has a toothy bonhomie and offers folksy feel-good sermons. His feisty blonde wife, Victoria, has a rather more animal appeal as she struts the stage in her pencil skirt and stilettos, intoning breathily, "When you grow in love, you grow in *me*. Let it get deep. Deep in *you*. That love is growing—so PUSH into God more." Crikey.

Nobody there wants to talk about climate change. The Osteens have no desire for an interview despite repeated attempts to get one. When I approach people after the service, many turn away and refuse to talk at all. Others claim ignorance or indifference.

Bob and Michelle from Nashville, though, are trapped alongside me in the pew, palms outstretched to absorb the blessings raining down on them. What do they think? Michelle turns away, unwilling to even discuss it. Bob reckons it's all a natural cycle, but he's sure God is in control. Later on I complain that it's freezing in the air-conditioned basketball stadium that passes for a church. "Yup," he says chuckling, "not much global warming in here."

The question on my mind—a reasonable one really—is to ask what Lakewood might have that the world's greatest crisis does not. Every week Lakewood Church achieves a level of mass mobilization that climate change activists can only dream about. Consider it this way: In February 2013, sixty environmental organizations pulled out the stops to mobilize forty-five thousand people for the largest-ever climate change rally in Washington, D.C. That week, just as many people came to this one church. And just as many came the next. Six times more people will watch this service on television and on the Internet than watched *An Inconvenient Truth* in U.S. cinemas.

If climate change campaigners complain about the lack of foundation funding or media coverage, they should try running an evangelical church. Churches generate their own media, raise their own money, publish their own books, and sell themselves entirely through the quality of the experience they offer converts. They are, as it were, real-time experiments in what moves, excites, and persuades people.

Ara Norenzayan, a social psychologist at the University of British Columbia, is determined to identify the winning psychological qualities that have created the world's dominant religions. After all, he tells me, there are ten thousand religions in the world, so there must be strong reasons why two-thirds of people have come to follow just three of them: Christianity, Hinduism, and Islam. These are, he argues, "the descendants of just a few outlier religious movements that have won in the cultural marketplace through two thousand years of successful experimentation."

Norenzayan is something of an outlier himself, exploring areas that other psychologists consistently ignore. He was one of the researchers who created the acronym WEIRD (western, educated, industrialized, rich, democratic) and has concluded that these same inward-looking assumptions have led psychologists to seriously underestimate the relevance of religion. He observes that experimental psychologists look around their small subculture and say, 'No one who is important to me is religious, so this must not be very important.'"

He strongly agrees when I suggest to him that the same criticism could be leveled at the climate change movement. "These people are ignoring the largest social movements in the world and the ones that have proven time and again to have the power to galvanize people into action," he says.

So what, I ask him, could the climate movement learn from his work on the psychology of religion? He thinks for a moment, and his answer is fascinating.

"From a WEIRD perspective," he says, "climate change appears to be hopeless because people will never be prepared to make a sacrifice because of the rational calculation. But this is not the case in religions, which contain sacred values that are so fundamental that they are entirely nonnegotiable. They cannot be bought or sold, and people will make any sacrifice to defend them."

Sacred values are not just about religion. Brain scans have found that the parts of the brain associated with sacred values are those associated with other moral choices. Sacred values are embedded throughout our culture—the defense of our children is a sacred value and we would not sell them at any price. Torture is considered to be wrong and is not subject to any temporal discounting—it will be just as wrong ten years from now as it is now. National parks are a sacred value to Americans—you could never sell Yellowstone.

For Norenzayan, a radical solution would require turning action on climate change into a non-negotiable sacred value. But could you mobilize sacred values without a religion? Absolutely he says—and in any case, a religion "is not like a thing; it's an assembly of features that become a group called religion. You could co-opt these successful qualities and use them in other contexts." His view echoes the work of the American sociologist Robert Bellah who argued that religion "is transmitted more by narrative, image, and enactment than through definitions and logical demonstration."

So, what are the features of the great religions and how might they be mobilized to create sacred values around climate change?

First—and I do not wish to be an apologist for the violence and coercion that often accompanied this process—they have all invested heavily in gaining new audiences through missionary outreach and proselytizing. Much of the growth of Mormonism is due to the high status given to missionary service. Churches have constantly experimented with ways to engage new cultures. Consider, for example, how Catholic missionaries adopted different tactics for working in China. The Franciscans charged in, declaring, "Here is the new God." The Jesuits, under the instructions of their leader, Matteo Ricci, wore Chinese robes, adopted the Chinese language, and avoided all contact with Europeans.

As religions recruited new members, they developed institutions to maintain a community of shared belief through ritual and shared worship. For the pioneering sociologist Émile Durkheim, religion was not just a social creation; it was, he said, society made divine. The reward for belief comes from belonging to the community of belief—and the cost of disbelief is social rejection.

Lakewood Church, by any standards a roaring success in the cultural marketplace, is fueled by the irresistible enthusiasm of its mass gatherings. Its critics, of which there are many among traditional Christians, see it as being little more than a weekly rock concert. But it is more intelligent than that. Pastor Joel Osteen focuses on relevance—giving people something to take away. He preaches around simple themes that are directly relevant to people's lives. Above all, he is upbeat: talking about self-esteem, confidence, and, taking a line from Jesus, how you can "become what you believe." Go higher in life, he urges, rise above your obstacles, live in health, abundance, healing, and victory. And this is why Bob from Nashville loves him so much—"you always feel so much better afterward," he says.

Receiving God's blessing and feeling good need not exclude talk of environmental responsibility or climate change. Northland Church in Longwood, Florida, approaches the scale and showmanship of Lakewood while embracing the message of climate change and caring for God's creation. Under the leadership of its charismatic pastor, Joel Hunter, Northland has grown into one of the thirty largest churches in the country. It is unusual for its experimentation with new communications technologies. Hunter describes his new forty-two-million-dollar church as a "communications device with a sanctuary attached," which enables Sunday services to be beamed to a live congregation of more than fifteen thousand people in three churches and services held in people's homes.

Hunter is warm, considerate, and funny—with his silver hair and broad smile, he looks rather like Jack Palance. I can fully understand why President Obama is glad to have Hunter as a friend and spiritual adviser.

This does not make Hunter into a liberal by any stretch. In his book *A New Kind of Conservative*, he outlines the central authority of the Bible against gay marriage and calls for personal responsibility and smaller government. Hunter likes to describe himself as an independent, deciding on his position issue by issue.

This independence has led to some predictable fire from the Florida far right for his open partnership with Muslim preachers. But that is nothing compared with the attacks he receives for his belief in climate change. The hate mail has now calmed down, but while he is rarely accused of being a "tool of the devil" anymore, people still take him aside to say, "I think you are good man, and you mean well, but *they've got you.*"

Hunter was introduced to the subject by his fellow evangelist Richard Cizik—"my friends are always getting me into trouble," he says—who had attended the 2002 conference organized by Sir John Houghton in Oxford. Cizik describes his experience in Oxford as a conversion experience: "I had, as John Wesley would say, a warming of my heart, a change that only God could do, like Paul's conversion in which he fell off a donkey on the road to Damascus."

Hunter also describes his belief in climate change as a religious conversion. He quotes from the Gospel of John, where Jesus says, "You must be born again. The wind blows wherever it pleases. You hear its sound, but you cannot tell where it comes from or where it is going." You are being called to live your life by a different standard.

Always on the lookout for new narratives, I invite Hunter to explore the ways that his church identifies and nurtures belief in Christ and whether these might help us build wider acceptance of climate change. Three key concepts emerge that are directly relevant to climate change.

First, that belief is held socially and is shared through testimony and witnessing with your peers and community. Hunter describes this as "the huge one": "You have the fellowship of fellow believers. That is the encouragement that we need, to be around people who have the same interests, the same goals, the same values as ourselves."

The church then becomes a safe place to admit to personal problems and struggles with belief and doubt. "We have to make an environment— excuse the pun—where we fully acknowledge that everybody is going to have doubts and struggles, and everybody is going to need encouragement. We see if we can help with that and we walk through that together."

Second, that people can be brought to a commitment at a moment of choice. In the Bible, people are offered choices: As God says to the people of Israel, "I have set before you life and prosperity, death and destruction, blessings and curses . . . Now choose life, so that you and your children may live." Evangelism seeks to generate a moment when people actively choose to commit themselves to their faith in a public context that sets a social norm for others. The outreach crusades of the great evangelist Billy Graham would culminate in an altar call in which new people were invited to step forward to receive a special blessing. It's a simple but very effective device to break the bystander effect. As Hunter says, "Even if you are a little tentative, you see all these people going forward and you think, 'I have nothing to fear. I will be with them. I'm not left to my own devices. I want to join the movement.'"

And, Hunter need not add, the altar call is also a point at which the church can identify potential new members and then welcome and support them. Evangelical outreach, such as the hugely successful "I Found It" advertising campaign in the 1970s, always directs people to make personal contact through their local church. Environmental outreach around climate change, on the other hand, invariably directs people to websites and places where they can find more information.

Third, Hunter argues that belief in climate change can be understood as a personal revelation. Moments of personal revelation are a universal human experience reported by around three-quarters of people, regardless of their culture or religion. In 1969, more than seven thousand people replied to a small advertisement placed in British newspapers inviting them to share their "experience of a presence or power which is different from your everyday self." They described their experience as joyous, sometime frightening, and always "ineffable" and "unknowable." Although these are often called religious experiences, only a quarter of the respondents use the word *God*.

Professor Brian Hoskins, the director of the Grantham Institute for Climate Change in London, is unusual in his recognition that scientific information requires this transformative moment. "Often what we do is provide the landscape in which Saint Paul can have his moment. I don't believe these come from nowhere; they come from all the information around and then it clicks for someone. We [as scientists] are creating the ether in which people can have that illumination."

Lynda Gratton, a chair of the World Economic Forum, reports that the most ambitious sustainability programs in the business world invariably stem from the transformative inner experience of a single influential individual. Jochen Zeitz, the former chairman of Puma, says that his stay in a Benedictine monastery inspired him to develop a valuation of environmental impact in his bookkeeping. H. Lee Scott, the former CEO of Walmart, reportedly had a climate change "epiphany" on a field trip in New Hampshire to learn about the impacts of global warming on maple trees.

Eamon Ryan, the former environment minister in the Irish government, told me of how an ecology course at his Jesuit school, which started as "a chance to lark about and smoke behind trees," became his personal "epiphany on the interconnectedness of us and nature." His language is itself a reflection of the teaching of Ignatius Loyola, the founder of the Jesuit movement, who had found his own calling sitting by a river.

Bob Inglis, former Republican representative for South Carolina, asks how we can enable this kind of life-changing conversion. Drawing on the church experience, he formulates it this way to me: "You go to them with a credible messenger and you affirm their truth. And you help them to see that this fits within their story and you honor them by being there with them. Then you can get conversions."

Conversions? Affirm? Witness? Epiphany? These words never appear in the discussions of how we mobilize action on climate change. Acceptance of climate change is assumed to be transferred, as though through osmosis, by reading a book or watching a documentary. When it is acquired, it is assumed, like the data that it is based on, to be solid and unshakeable. Because there is no recognition of climate change conviction, there is no language of climate change doubt, no one is offering to give us encouragement or to help us to "walk though that together." There is no defense against backsliding and denial, and there is no mechanism for coping with grief.

And so, outside the circles of dedicated environmental activists, there is no community of belief. There is no social mechanism for sharing it, least of all witnessing it. People deal with their hopes and fears in isolation, constrained by the socially policed silence and given no encouragement other than a few energy-saving consumer choices. If Christianity were promoted like climate change, it would amount to no more than reading a Gideon's Bible in a motel chalet and trying to be nice to people. The critics are right in this regard—if climate change really were a religion, it would be a wretched one, offering guilt and blame and fear but with no recourse to salvation or forgiveness.

Guilt is a word that appears all the time around climate change. "Is it just me, or does everyone else feel guilty for being alive too?" wrote Jeremy Burgess in an opinion piece in *New Scientist*. What is missing, Burgess noted, is forgiveness, and failing this, "we can only look forward to punishment."

Sally Weintrobe, a psychotherapist, agrees. Without forgiveness, she writes, our feelings about climate change will "become stuck in a climate of hatred, bitter recrimination and relentlessness, easily feeling harshly judged and not moving towards accepting the reality of the loss."

Mechanisms for personal forgiveness are a critical component of the Abrahamic religions. In the Christian faith it is the power of God to forgive that leaves the door open for personal change. In Judaism the ritual of Al Chet, held on the eve of the Day of Atonement, Yom Kippur, enables the recital of sins against oneself, others, and God. Most relevant to climate change, the ritual explicitly regards inaction and silence as the moral equivalents of active sin.

The climate change narrative contains no language of forgiveness. It requires people to accept their entire guilt and responsibility with no option for a new beginning. Not surprisingly, what happens is that people either reject the entire moralistic package or generate their own self-forgiveness through ingenious moral licensing.

Fred Luskin, the director of the Stanford University Forgiveness Projects, is at the center of the booming area of forgiveness research. He tells me that the number of published studies has quadrupled in just ten years, though not one of them has been concerned with forgiveness and climate change. He agrees that this is a major weakness, especially, he suggests, as climate impacts increase and "there will be a frantic rush to punish, to assign blame, to limit freedoms, and to set up good guys versus bad guys."

This is already happening in international negotiations, where the unresolved responsibility for past emissions continues to prevent agreement on a shared approach to future action. According to Luskin, forgiveness is not about pardoning, or excusing; it is "a process of transforming the continuing and destructive feelings of guilt, blame, and anger into positive emotions such as empathy and reconstruction." The absence of a narrative of forgiveness cuts off many of the options for a constructive resolution.

I found talking with Hunter and other evangelicals invigorating. Putting aside the immediate context of evangelical Christianity—and here I will openly admit that I respect but do not share their religious faith—they outline a vocabulary and methodology for overcoming our cognitive obstacles that is absent from discussions around climate change.

In this book I have shown that scientific data, although undoubtedly vital for alerting our rational brain to the existence of a threat, does not galvanize our emotional brain into action. Indeed, I have suggested, climate change contains enough inherent uncertainty and distance that we can quite deliberately keep what we know contained and detached from what we believe and what we do.

Learning from religions, I suggest that we could find a different approach to climate change that recognizes the importance of *conviction*: the point at which the rational crosses into the emotional, the head into the heart, and we can say, "I've heard enough, I've seen enough—now I am convinced."

Applied to climate change we could accept that this is a process of steadily growing awareness, though it may also progress through personal revelation or moments of informed choice and public commitment. Conviction need not remove questioning and doubt—and it is essential that climate science is never above challenge—and these uncertainties, anxieties, and fears need to be openly recognized within a supportive community of shared conviction.

Finally we could learn to find ways to address the feelings of blame and guilt that lead people to ignore or deny the issue, by enabling people to admit to their failings, to be forgiven, and to aim higher. By concentrating on universal and non-negotiable "sacred values," we could sidestep the arid cost-benefit calculations which encourage us to pass the costs onto future generations.

These ideas are not unique to religions, and can be found in every successful social and political movement in history. We already know how to overcome the cognitive challenges that make it possible for us to ignore climate change. The lessons are all there, if we choose to heed them.

41

Why We Are Wired to Ignore Climate Change . . . And Why We Are Wired to Take Action

THROUGH OUR LONG EVOLUTION, WE have inherited fundamental and universal cognitive wiring that shapes the way that we see the world and interpret threats and that motivates us to act on them. Without doubt, climate change has qualities that play poorly to these innate tendencies. It is complex, unfamiliar, slow moving, invisible, and intergenerational. Of all the possible combinations of loss and gain, climate change contains the most challenging: requiring certain short-term loss in order to mitigate against an uncertain longer-term loss.

Climate change also challenges and reverses some deeply held assumptions. We are told that the way of life that we associate with our comfort and the protection of our families is now a menace; that gases we have believed to be benign are now poisonous; that our familiar environment is becoming dangerous and uncertain.

Our social intelligence is well attuned to keeping track of debts and favors, and ensuring equitable distribution of gains and losses. Climate change poses a major challenge here too, with all solutions requiring that rival social groups agree on a distribution of losses and thereafter the allocation of a greatly diminished shared atmospheric commons.

We are best prepared to anticipate threats from other humans. We are inordinately skilled at identifying social allies and enemies, identifying the social cues that define loyalty to our group and that identify the members of rival out-groups. Climate change is immensely challenging in terms of these categorizations. It is not caused by an external enemy with obvious intention to cause harm. It therefore tends to be fitted around existing enemies and their perceived intentions: a rival super-power, big government, intellectual elites, liberal environmentalists, fossil fuel corporations, lobbyists, right-wing think tanks, or social failings such as overconsumption, overpopulation, or selfishness.

Worse still, and unique among major threats, we all contribute directly through our own emissions and are therefore personally responsible for the ever-increasing costs for ourselves, our in-group, and our children and descendants. This moral challenge, combined with a sense of the relative powerlessness of individual action, helps mobilize a well-ingrained set of defense mechanisms that enables us to ignore the problem—both through personal disavowal and through socially constructed silence.

There is a fundamental division, embedded in the physical structure of our brain, into the analytic and the experiential processing systems—what I have called the rational brain and the emotional brain. The two brains work together on complex tasks, but the engagement of the emotional brain is critical for galvanizing action, especially at a social level. The differences between our rational and our emotional processing systems express themselves in a constant tension between the overly rational presentation of climate science and its translation by campaigners into emotionally appealing narratives.

The cognitive systems require that complex issues be converted into narratives which become the primary medium by which the issue and the social cues that guide attention are transmitted between people. Meaning is therefore created by the way we talk about it (or, I have suggested, the ways that we choose not to talk about it).

Stories and narratives have universal qualities, and we squeeze new information into these standard story patterns. We then justify these stories with reference to available recent experience—usually itself in the form of a socially generated story.

Climate change is, I suggest, exceptionally *multivalent*. It lends itself to multiple interpretations of causality, timing, and impact. This leaves it extremely vulnerable to our innate disposition to select or adapt information so that it confirms our preexisting assumptions—biased assimilation and confirmation bias. If climate change can be interpreted in any number of ways, it is therefore prone to being interpreted in the way that we choose.

These constructed narratives therefore contain the final reason why we can ignore climate change: they become so culturally specific that people who do not identify with their values can reject the issue they explain.

The narratives formed by the early adopters of the issue came to dominate and frame all subsequent discussion. The early focus on tailpipe emissions rather than wellhead production became a meta-frame that influenced all subsequent narratives concerning the definition of the problem, moral responsibility, and policy solutions.

As the issue matured, deniers became louder and stronger and created their own narratives that came to "pollute" the discourse. These built on and reacted to the existing narratives, often adopting and reworking their frames, to create compelling stories in which familiar enemies were motivated by self-interest to cause intentional harm.

As these narratives became repeated and shared within peer groups, they came to constitute a social proof. These reinforced the other social cues coming from the media and political elites. As the issue developed, these cues accumulated and powerful social feedbacks tended to amplify them, leading people to overestimate the consensus within their own social group and to alter or suppress their own opinion if it did not conform.

What is more, we are all active participants in this process, developing personal narratives that help us to manage the anxiety, moral challenge, and required sacrifices inherent in climate change by choosing to make it yet more distant, less certain, more hopeless, or less relevant to our own values. We even interpret the wider social norms to select the social cues that best reinforce our chosen position. That is to say that, even with the best intentions, we cannot help setting up narratives that are designed to fail against the very biases they are supposed to overcome.

There is, then, no single factor that leads people to ignore climate change. Anyone who suggests that there is will, inevitably, be fulfilling the wicked prophecy and defining the problem to support that conclusion. Rather, there is a set of interrelated negotiations between our personal self-interest and our social identity, in which we actively participate to shape climate change in ways that enable us to avoid it.

The bottom line is that we do not accept climate change because we wish to avoid the anxiety it generates and the deep changes it requires. In this regard, it is not unlike any other major threat. However, because it carries none of the clear markers that would normally lead our brains to overrule our short-term interests, we actively conspire with each other, and mobilize our own biases to keep it perpetually in the background.

. . . And Why We Are Wired To Take Action

Even with our limitations, humans can accept, understand, and take action on anything. We have immense capacity for pro-social, supportive, and altruistic behavior. Climate change is entirely within our capacity for change. It is challenging, but far from impossible.

Beyond immediate personal threats, we have no instinct stronger than the drive to defend the interests of our own descendants and social group. Climate change is not a minor inconvenience—even though some narratives shape it as such. It is an existential threat on a scale equaled only by nuclear war. It contains threats at every level: to our sense of place, our identity, our way of life, our expectations of the future, and our deepest instincts that lead us to protect our children and defend our tribe.

Nothing is contained within climate change that we are incapable of dealing with. Even though it presents itself in the form of a future threat, we have the capacity to anticipate threats, by giving them the narrative and cultural form that engages our emotional brain and by creating social institutions that sustain our response. We have a strong drive toward such collective enterprises, for they are one of the means by which we cope with the fear of our own mortality.

We also have a virtually unlimited capacity to accept things that might otherwise prove to be cognitively challenging once they are supported within a culture of shared conviction, reinforced through social norms, and conveyed in narratives that speak to our "sacred values." These could just as readily lead us to action as lead us to inaction.

There is no single pathway from information to conviction. The cultural feedbacks that lead climate change to become more distant, uncertain, or hopeless could equally well work the other way by creating a social proof and legitimacy around accepting and taking action. The personal reward for action would then come from an intensified sense of belonging and the satisfaction that comes from contributing to a shared project. Climate change is the one issue that could bring us together and enable us to overcome our historic divisions. This, rather than the self-interest contained in the economic arguments, is the real reward of taking action.

The final proof that we are not inherently "wired" to ignore climate change—which should be self-evident—is that the majority of people, across the world, already accept that it is a major threat and might be prepared to support the necessary changes. They currently feel isolated and powerless, but could readily be mobilized if their concerns and hopes became validated within a community of shared conviction and purpose. Human history provides so many examples of social movements that have overcome apparently impossible obstacles that we know that we should be capable of meeting this challenge, providing that we move decisively.

But this is just one of the many pathways that are opening up in front of us. Climate change is not a static issue, and extreme weather events of entirely unprecedented scale and duration will continue to build. These events now occur within a cultural and political environment that has been thoroughly primed with socially charged beliefs. The critical questions for the future are how the increasing personal experience of extreme weather will interact with these existing narratives, and whether the result will be an increase or a decrease in our acceptance that our own behavior is their underlying cause.

42

In a Nutshell

Some Personal and Highly Biased Ideas for Digging Our Way Out of This Hole

CLIMATE CHANGE IS A SCIENTIFIC fact. Scientists have become so bruised by their political battles that they have come to use much weaker language, declaring that climate change is "very likely" or "unequivocal." Let's just call it a fact, because that is what it is. There is plenty of uncertainty around how the climate is responding to these enormous changes, but being uncertain is not the same as being unsure.

Scientists are remarkably sure that climate change is bringing major impacts—they simply cannot with absolute certainty disentangle the web of cause and effect. The word *certain* is one of those many false friends of words that scientists use in a particular and unusual meaning. In regard to climate change, we are frequently divided by our common language.

Our psychological obstacles are also a scientific fact. The large body of rigorous research-based evidence suggests that climate change struggles to overcome numerous biases against threats that appear to be distant in time and place. We need to make these explicit and recognize that many may be subconscious.

To create proximity we need to **EMPHASIZE THAT CLIMATE CHANGE IS HAPPENING HERE AND NOW**. In particular, we should **BE WARY OF CREATING DISTANCE** by framing climate change as a future threat for people far away and, especially, as a threat for non-humans, however cute they might be.

Our sense of loss looks backward rather than forward, and research suggests that people are more motivated to restore lost environmental quality than improve current environmental quality. There is therefore a potential to express climate change as an opportunity to **RESTORE PAST LOSS**, whether it is social (lost community, values, purpose) or environmental (lost ecosystems, species, or beauty). The rapidly growing movement for the rewilding of degraded landscapes is an interesting response to the uncertainties of future loss.

We are very well adapted to respond to immediate threats but slow to accommodate moving change. Climate change is a process, not an event, so it requires that we **RECOGNIZE MOMENTS OF PROXIMITY** that can demand attention. These may be moments of political decision making, collective action, or generated conflict. In my view, the Keystone XL pipeline is a legitimate attempt to create a historic moment. Those critics who argue that the pipeline will only ever be a small part of overall U.S. emissions are missing the point. Their complaint is like saying that the locations of seats at the lunch counter of the Greensboro Woolworth's or on the Montgomery buses were trifling examples of racial segregation. Sometimes the act of **CREATING THE SYMBOLIC MOMENT** is far more important than its overall relevance.

Extreme weather events create a moment of proximity and heightened awareness, but also of the increased in-group loyalty and anxiety that can readily exclude consideration of climate change. Even when confronted with direct evidence of climate extremes, the main influence on people's attitudes will still be the views of the people they know and trust.

The interference of outsiders will very likely be counterproductive in such situations, and the best option for building conviction lies with providing the information for trusted local communicators to **OPEN UP A CONVERSATION ABOUT LONG-TERM PREPAREDNESS**. Preparedness and adaptation are routes for people to accept that climate change is real and already under way—and, as I have shown, it is possible to build a discussion around these topics even when it is politically taboo to talk about the wider issues.

However, these approaches will always be specific to each context. Whatever the findings of psychology experiments with their WEIRD experimental subjects, we need to remember that not everyone wants to protect the status quo, especially if they are already struggling against economic and social injustice. So we need a **NARRATIVE OF POSITIVE CHANGE**, in which our adaptation to climate change does not just protect what is already here but also creates a more just and equitable world.

Climate change is a narrative, shaped through social negotiations and transmitted between peers. People form their response to the narratives, not the science, and so it always needs to **FOLLOW NARRATIVE RULES, WITH RECOGNIZEABLE ACTORS, MOTIVES, CAUSES, AND EFFECTS**. People will be inclined to follow the most compelling narrative, so be careful: **DON'T LET THE NARRATIVE TAKE OVER** the way we think or talk about it.

We interpret climate change through frames, which focus our attention but limit our understanding—they allow us to exclude or ignore meanings that lie outside the frame. Most of the factors that enable us to ignore climate change derive from attempts to limit its meaning; that it is an *environmental* issue, *a threat* or an *opportunity* (but not both), a *wellhead problem* or a *tailpipe problem* (but not both). So, **RESIST SIMPLE FRAMINGS** and **BE OPEN TO NEW MEANINGS**.

Because climate change is a wicked problem, it can easily become defined entirely by its own framings and the solutions we propose, and policy makers can easily become locked into the simple one-off solutions that solve tamer problems. We all need to **ENSURE THAT A WIDE RANGE OF SOLUTIONS IS CONSTANTLY UNDER REVIEW**—a process that planners call iterative risk management.

Frames define battlegrounds, and so limited frames can lead to false debates. Arguments that renewable energy brings greater energy security encourage the expansion of domestic fossil fuels. Arguments that the low-carbon economy will bring jobs become vulnerable to evidence that the high-carbon economy might bring more jobs. As the cognitive linguist George Lakoff says, **NEVER ACCEPT YOUR OPPONENT'S FRAMES**— "don't negate them, or repeat them, or structure your arguments to counter them."

The presence of enemies with the intention to do harm engages our moral brain and energizes our outrage. However, climate change lacks clear enemies: We all contribute to this problem and all stand to suffer its impacts. This is an incomplete and uncompelling narrative, and activists on all sides seek enemies that can fill these missing roles of good against evil, David against Goliath, might against right.

We need major change, and change requires social movements. Some argue that movements need enemies, and this may well be true for generating rapid change. However, there is also a price to pay. This is an in-group, out-group game, so **BE CAREFUL THAT ENEMY NARRATIVES DO NOT FUEL DIVISION** or agitate deep-rooted and distracting animosities at a time when we need to be finding common purpose. My view is that campaign narratives could experiment more with alternative narrative traditions, for example **CREATE A HEROIC QUEST** in which the enemy may be our internal weaknesses rather than an outside group.

Overall, we need to **BUILD A NARRATIVE OF COOPERATION** that can bring people together around a common cause. This should **STRESS COOPERATION NOT UNITY**—we do not have to become the same people, and conservatives in particular require well-defined differences rather than a merger. **ACCEPT THE SPECTRUM OF APPROACHES** with radical protesters, lobbyists, policy makers, and multiple different sectors, all pushing in the same direction if not with the same detailed objectives.

In the way that we tell the climate change story, we need to **BE HONEST ABOUT THE DANGER**—but remember that this will only motivate people if they hear it from trusted communicators and can see opportunities for action and change. **ENCOURAGE POSITIVE VISIONS**, but remember that these may carry social cues that may repel others. The bright side technocratic future vision, for example, is elitist and materialistic, and alienates those who already feel disenfranchised.

When people say that climate change requires a values change, they invariably mean that other people need to change to *their* values. In fact we *all* hold the right values, and humans have an extraordinary capacity to empathize and care about the welfare of others. The problem is that we have not all engaged the right values with this issue. The challenge is how to best **ACTIVATE COOPERATIVE VALUES RATHER THAN COMPETITIVE VALUES. STRESS WHAT WE HAVE IN COMMON**: a better life for our children, health, security, thriving communities.

By contrast, attempts to motivate people though appeals to personal self-interest are unlikely to be successful. Contrary to the assumptions of conventional communications, extensive research confirms that people are poorly motivated by money. Money is important, but it is a proxy for other ends: security, caring for your family, and social identity, which could be addressed in other ways. It is far more effective to **RELATE**

SOLUTIONS TO CLIMATE CHANGE TO THE SOURCES OF HAPPINESS, and the connections we feel with our friends, neighbors, and colleagues.

People are best motivated when an action reinforces their identity and sense of belonging to their social group. **EMPHASIZE THAT ACTION ON CLIMATE CHANGE MAKES US PROUD TO BE WHO WE ARE**, and reinforce this with the *social cues* and *social proof* that people like ourselves are seen as concerned and taking action. Most communication around climate change and low-carbon behaviors is anti-replicating, based around loneliness, isolation, and despair. So **ENABLE COMMUNICATIONS WITH BUILT-IN INTERACTION** that can be passed between peers and create visible social norms. We need to stop regarding climate change as an isolated intellectual exercise and **CREATE COMMUNITIES OF SHARED CONVICTION** within which people can share their doubts and fears and draw on the strength of shared commitment.

Climate change is a science *and* a conviction. Following the division built into our own brains between our rational and emotional processing systems, it is entirely possible to know about climate change and yet not to fully believe in it. Conviction is the critical process by which we incorporate climate change into our moral framework and accept the need for action.

A conviction is not a blind faith: We should continue to **KEEP AN OPEN MIND**. There is an excessive level of closed-mindedness on all sides, and two-thirds of people say that they will never change their minds about climate change. Because climate change is ambiguous and multivalent, it is open to multiple interpretations. So **BE ALERT TO YOUR OWN BIAS** and to your own innate tendency to select the information that confirms your existing views.

REMEMBER THAT EXPERTS CAN ALSO BE BIASED by their own specialism or worldview. Clever people indulge in clever confirmation bias. Experts are human too and are also coping with their own internal conflicts, which they may be projecting onto the way that they interpret climate change. So always **SEEK OUT A WIDE RANGE OF VIEWS**.

Listen to people who disagree with you, and recognize that they can sometimes be a source of insight and alert you to your own bias. **DEBATE IS USEFUL** so **LEARN FROM YOUR CRITICS**.

And, for the benefit of conservatives and skeptics, I would add that you, too, should listen to the other side and **RESPECT**

ENVIRONMENTALISTS, who have worked for three decades to keep this issue alive. If you do not like what they say, then you should become more involved in building positive solutions around your values rather than fighting a losing battle to undermine the science.

We should **BE PREPARED TO LEARN FROM RELIGIONS** and the thousands of years of experience they have in creating methods to sustain socially held belief. This does not mean that climate change is a religion, any more than a declared belief in the right to personal freedom, sound finance, or the strength of the military are religions—these are statements of commitment to personally held ideals (taken, as it happens, from Republican presidents).

Learning from religions, we can **PRESENT CLIMATE CHANGE AS A JOURNEY OF CONVICTION** which will contain periods of doubt and uncertainty as well as moments of personal revelation and sudden awareness. Encourage people to explain, in their own words, these moments and the *process* by which they came to terms with the science, recognizing that conviction is sometimes hard to maintain and needs to be reaffirmed.

We should also **CREATE MOMENTS OF COMMITMENT** and **FRAME CLIMATE CHANGE AS AN INFORMED CHOICE** between desirable and catastrophic outcomes, in which people can understand that inaction is itself a choice in favor of severe climate change.

To break through the self-interest of our cognitive biases, and fully activate our emotional brain, we need to **INVOKE THE NONNEGOTIABLE SACRED VALUES** that would enable people to make short-term sacrifices for the long-term collective good—for example, values that prohibit destroying a precious cultural asset, inflicting harm on the weak or innocent, abusing God's creation, and being cruel to our parents or children.

In the formation of conviction, trust is more important than information. Communicators, especially scientists, should learn to **EMPHASIZE THE QUALITIES THAT CREATE TRUST** (their independence, values, accountability) and especially **TELL PERSONAL STORIES**. Communicators should talk about their personal journey, especially if they have come to their conviction from a position of doubt. They should **BE EMOTIONALLY HONEST**, talking openly about their hopes, fear, and anxieties.

Moral consistency is especially important for trust. If you wish to communicate climate change, you need to **RECOGNIZE THE ROLE OF YOUR OWN EMISSIONS**, not least because a high-emission lifestyle will inevitably corrupt your own judgment, and you should share your own struggle and success in reducing them

Campaigners and politicians love to fantasize that a huge top-down communications projects will finally knock it into people's heads. They are unlikely to work. Instead we need to **ENABLE FRESH, REAL VOICES**, and not depend on the glib slogans of advertising agencies. And this means that the people who currently communicate climate change, especially environmentalists, must be prepared to **BACK OFF AND ENCOURAGE NEW COMMUNICATORS**—not as the guests on their podium but as new speakers in their own right.

Actually, let's go a step further. Climate change does not belong to environmentalists and is not even environmental. Of course, it includes environmental concerns and impacts, but it is so much bigger than that. As soon as we label it, we restrict our understanding of it. Obviously, environmentalists can talk about it however they like in their own networks, but for wider presentation and to the media, I plead, **DROP THE ECO-STUFF**, especially polar bears, saving the planet, and any other language that stakes out climate change as the exclusive cultural domain of environmentalism.

Above all, it is critical that we **CLOSE THE PARTISAN GAP** between left and right by opening up climate change to conservative framings and ownership. This should start with **AFFIRMING WIDER VALUES**, which, it is well established experimentally, makes people far more willing to accept information that challenges their worldview. This requires communicators to reverse the normal flow that converts the science into people's values and begin by understanding and validating their values first and then come up with the ways that climate change can speak to those values.

Testing suggests that new framings of values could include respect for authority, personal responsibility, and loyalty to one's community and nation, avoiding intergenerational debt, and reducing societal dysfunction. I warn environmental liberals that the measure of success will inevitably be the emergence of some new ways of talking that you find unpleasant. Similarly, **NEVER ASSUME THAT WHAT WORKS FOR YOU WILL WORK FOR OTHERS**. Indeed, the fact that you strongly like something may well be an indication that people with other values will hate it.

We also need to **BE HONEST—THIS IS TOUGH**. Psychotherapists argue that the real challenge is that climate change generates strong feelings that can, unless recognized, lead us to disavowal and outright denial. We need to **RECOGNIZE PEOPLE'S FEELINGS OF GRIEF AND ANXIETY**, and acknowledge and provide space for contradiction, ambivalence, loss, and mourning.

The starting point could be providing the space for people to openly acknowledge their feelings and share them. We need to **MOURN WHAT IS LOST, VALUE WHAT REMAINS**. And not just the natural world; we need to **MOURN THE END OF THE FOSSIL FUELS AGE**, which, for all of its dirt and danger, was also exceptionally affluent, mobile, and exciting. The low-carbon world will have new pleasures, but no longer the sweet roar of the Ford Mustang V8.

We should all **BE GLAD TO BE A POLLYANNA**. She has become synonymous with dim-witted optimism, but in the original books by Eleanor H. Porter, the character is clearly shown to be coping with immense grief and suffering through her gratitude for what she does have—her friends, community, and the joy of being alive.

What is clear is that this is a fast-moving issue and everything will change. At present, climate change exists largely as a narrative of anticipation shaped by familiar experience and existing frames. But momentous shifts are under way in the world's climate systems and carbon cycles, which will, within a single lifetime, make climate change entirely real, salient, and unavoidable. This will be a new world in which past certainties will disappear and our inbuilt social and psychological biases will become increasingly influential on our judgment.

This is why current responses are so important. **REMEMBER THAT HOW WE RESPOND NOW WILL PROVIDE THE TEMPLATE FOR FUTURE RESPONSES**. Acceptance, compassion, cooperation, and empathy will produce very different outcomes than aggression, competition, blame, and denial. We hold both futures within ourselves and, as we choose whether and how to think about climate change, we are choosing how we will think about ourselves and the new world we are creating.

Four Degrees

Why This Book Is Important

IN THE INTRODUCTION TO THIS book I pledged that it would not contain information on the impacts of climate change until its final chapter. Later I discussed how scientists struggled to maintain their composure in the face of the information that they held. In particular I mentioned their anxiety that average global temperatures might rise over the threshold of 4 degrees Celsius (7.2 degrees Fahrenheit).

For many years their attention was focused on lower outcomes—especially around two degrees, the level that was adopted by policy makers, somewhat arbitrarily, as the boundary level for "dangerous" climate change. In recent years, though, scientists have become far more willing to warn that four degrees is the actual future we face. Professor Robert Watson, the co-chair of the IPCC, was the first to break ranks in 2008 when he publicly warned governments that they needed to develop adaptation plans for four degrees. The following year international experts met for the first time to present detailed scenarios at the "4 Degrees and Beyond" conference at Oxford University. By 2013, there was sufficient agreement that Mark Maslin, professor of climatology at University College London, could tell the Warsaw climate negotiations, "We are already planning for a 4°C world because that is where we are heading. I do not know of any scientists who do not believe that."

Four degrees is also increasingly on the minds of senior policy makers. The International Energy Agency reports that current emissions figures

put us on course for four degrees. In 2012 the World Bank, hardly a radical environmental organization, produced a major report with the title
"Why a 4°C Warmer World Must be Avoided." In his introduction the
Bank's president, Dr. Jim Yong Kim, said that he would ensure that "all
our work, all our thinking, is designed with the threat of a 4°C degree
world in mind."

So what does four degrees mean? Scientists, who are, as a group,
extremely wary of exaggeration, nonetheless keep using the same word:
catastrophe. Professor Steven Sherwood, a meteorologist at the University
of New South Wales, Australia, says that it would be "catastrophic,"
making life "difficult, if not impossible," in most of the tropics. Professor
Kevin Anderson, the former director of the Tyndall Centre for Climate
Change Research, says that it is hard to find *any* scientist who considers
four degrees "as anything other than catastrophic for both human society
and ecosystems."

I am going to resist the temptation to batter you with statistics—there
are some excellent detailed sources available online. Here instead are a
few snapshots of the four-degree world.

1. **Heatwaves**. In the words of the World Bank, there would be a
   "new class of heatwaves of magnitudes never experienced
   before"—indeed, with temperatures not seen on Earth in the
   past five million years. Four degrees will be a global average, so
   temperatures over large land masses will rise far more than this,
   by six degrees over North Africa, the Middle East, and the contig-
   uous United States. The warmest July in the Mediterranean
   region could be nine degrees Celsius warmer than today's warmest July.

2. **Extinctions**. Forty percent of plant and animal species will be at
   risk of extinction and the regional extinction of entire coral reef
   ecosystems would happen far earlier. Forests would be particularly vulnerable. A third of the Asian rainforests would be under
   threat and most of the Amazon would be at high risk of burning
   down.

3. **Food yields**. A three-degree rise causes all crops to experience a
   precipitous decline in their current growing regions. Overall
   yields could fall by a third in Africa. By some estimates temperature rises of over four degrees could reduce U.S. production of

corn, soybeans, and cotton by 63 to 82 percent. And there would be other pressures. In Africa and Australia 60 percent of current croplands would also be subject to extreme and recurrent droughts. These problems would be exacerbated by flooding, storms, and increased weed and pest invasions.

Other, equally catastrophic impacts follow close behind. Four degrees guarantees the total melting of the Greenland ice sheet and, most likely, the Western Antarctic ice sheet, raising sea levels by a combined thirty-two or more feet. The timescales are uncertain, but not the outcome: two thirds of the world's major cities and all of southern Bangladesh and Florida would end up underwater. Nor is there any guarantee that temperatures would level off at four degrees—at this level further powerful feedbacks and tipping points could lead temperatures to keep rising even further, to six and then eight degrees.

The research is developing and is still missing a strong sense of how these changes might interact with one another. What would be the combined effects of repeated droughts *and* storms *and* heatwaves *and* sea level rise? How would a world with nine billion people cope with such dramatic declines in the productivity of its main agricultural regions? What will happen to people in regions that are already marginal for human settlement when they become entirely uninhabitable?

Commenting on such interactions, Dr. Rachel Warren, a climate modeler at the Tyndall Centre for Climate Change Research, writes that "the limits for human and natural adaptation are likely to be exceeded." The World Bank echoes this when it concludes that there is "no certainty that adaptation might be possible." It is hard to comprehend what is meant by this abstract language, phrased, as so often, in the passive voice. Professor John Schellnhuber, one of the world's most influential climate scientists, is more direct: Speaking at a 2013 conference on the risks posed by a four-degree climate to Australia, he said that "the difference between two and four degrees is human civilization."

So when will we get there? Reviewing the current research, a British research team concluded that we could reach the four-degree point by the 2070s although, it noted, the 2060s are also possible.

However, the science around four degrees keeps moving—usually in the direction of greater pessimism. A recent paper from the University of New South Wales, Australia, argues that a decline of cloud cover in the tropics will accelerate warming so much that global temperatures could reach four degrees by the midcentury and, potentially, eight degrees by the end of the century. So much for climate change being a problem for future generations.

These predictions are couched in caveats and uncertainties, but these are usually a matter of the timeline rather than the outcome. The key variable, on which they all agree, is the level of emissions (especially those from burning fossil fuels), and the speed with which we reduce them.

And so, once again, we return to the overarching influence that our psychological response—our acceptance, avoidance, or denial—has in determining which path we will take. The shifting language of climate science reflects the growing evidence that our collective decision to ignore climate change commits us to a pathway along which we are rapidly losing any future options for control or choice. And this is why the recognition, understanding, and resolution of the questions explored in this book are so critically important.

References, Sources, and Further Reading

A FULL LIST OF LINE-BY-line references and sources is posted on the book website, www.climateconviction.org. There is a lively and continuing discussion and the references are regularly updated with new comments and clarifications.

From more than seven hundred sources, I have chosen a short selection of those I have found most useful and stimulating. Those indicated with a * can be found and downloaded on the Internet. I have shortened some of the titles for ease of searching.

Overviews on the Psychology of Climate Change
Clive Hamilton has given years of thought to the issues I cover and his book, *Requiem for a Species* (2010, Routledge), was a major inspiration. His insights can also be found online in his 2009 paper with Tim Kasser, *"Psychological Adaptation to the Threats and Stresses of a Four Degree World."

There are two major online summaries of climate change psychology. *The Psychology of Climate Change Communication: A Guide* by Sabine Marx is a very nicely produced overview. A more technical summary is *Psychology and Global Climate Change*, produced by an expert task force at the American Psychological Association.

Kari Norgaard's pioneering analysis of climate denial and silence in her book *Living in Denial* (2011, MIT Press) is very readable. She summarizes her ideas in an online report for the World Bank, *"Cognitive and Behavioral Challenges in Responding to Climate Change."

Another innovative analysis of the underlying psychology of climate change—this time from the psychoanalytic tradition—comes in a selection of essays, some highly insightful, in *Engaging with Climate Change: Psychoanalytic and Interdisciplinary Perspectives*, edited by Sally Weintrobe (2012, Routledge).

On cognitive psychology and decision making I strongly recommend Daniel Kahneman's *Thinking, Fast and Slow* (2011, Penguin). Eviatar Zerubavel's *The Elephant in the Room* (2007, OUP) is an excellent introduction to thinking about socially constructed silence. I have also been strongly influenced by the writings of Jonathan Haidt and Daniel Gilbert.

A summary of the research on narrative formation and the resistance of deeply held beliefs to challenge can be found in Stephan Lewandowsky and colleagues' little-known paper *"Misinformation and Its Correction: Continued Influence and Successful Debiasing." Despite its intimidating title, this is actually a very accessible and fascinating study. Dan Kahan's research and blog posts can be found on www.culturalcognition.net and a good starting point is his opinion piece for *Nature*, *"Why We Are Poles Apart on Climate Change."

In *Why We Disagree About Climate Change* (CUP, 2009) Mike Hulme discusses the "wicked" influence of cultural bias on our perceptions of risk and climate narratives, drawing on his twenty-five years of experience as a senior climate scientist, and, interestingly, his strong Christian faith.

There are good books on the psychology of risk by Dan Gardner and David Ropeik and, for a synthesis of the academic research, I can recommend *Risk: A Very Short Introduction* (2011, OUP). A summary of research on climate risk is contained in the paper *"Why Global Warming Does Not Scare Us (Yet)" by Elke Weber.

The Climate Issue and Arguments For and Against Action

There is so much strong writing on climate change, much of it online, that I cannot pick out any single source, though I am consistently stimulated by George Monbiot, Bill McKibben, Naomi Klein, Joe Romm, Dave Roberts, and Sharon Astyk. At the disaster end McKibben's 2012 *Rolling Stone* article, *"Global Warming's Terrifying New Math" is deservedly famous.

The arguments for a more positive technocratic message are presented in *Sizzle* by Futerra Communications, the materials of Sustainia, and the many publications of the Breakthrough Institute, starting with its deliberatively provocative report, *The Death of Environmentalism*.

Theda Skocpol's *Naming the Problem: What It Will Take to Counter

Extremism and Engage Americans in the Fight Against Global Warming is a challenging critique of the campaign to bring about climate change legislation. Skocpol does not pull any punches, which led to some lively online wrangling with climate activists.

A detailed analysis of the climate denial campaign as an ideological movement can be found online at www.desmogblog.com and in two good books: Naomi Oreskes's *Merchants of Doubt* (2012, Bloomsbury) and James Hoggan's *Climate Cover-Up* (2009, Greystone). Two Greenpeace reports expose Koch Industries: *Secretly Funding the Climate Denial Machine* and *Still Fueling Climate Denial.* For a discussion of the personalities of professional deniers, I recommend the very enjoyable *Washington Post* article *"The Tempest"* by Joel Achenbach.

For a better understanding of the moral universe of radical conservatism see *The Tea Party and the Remaking of Republican Conservatism* by Vanessa Williamson and others and COIN's report *A New Conversation with the Centre-Right about Climate Change.*

One of the very few detailed proposals for an international climate policy that controls wellhead production is contained in Oliver Tickell's important book *Kyoto2* (2008, Zed Books), based on the auction of permits to extract carbon.

I avoided discussion of the science itself, but, for my money, Mark Lynas's book *Six Degrees* (National Geographic, 2008) has still to be bested as a highly readable summary of future impacts. My main sources were the somewhat more opaque World Bank report *Turn Down the Heat: Why a 4°C Warmer World Must Be Avoided* and the online papers from the 2009 *"Four Degrees and Beyond"* conference.

There are numerous books criticizing climate science, usually very dull, but the most entertaining attack on climate change as a belief system is James Delingpole's *Watermelons: How Environmentalists Are Killing the Planet.* Delingpole has no interest in science or evidence, so this is just pure narrative-driven bile.

Public Opinion

Matthew Nisbet and Teresa Myers conducted an exhaustive overview of polling from 1987–2007, charting the rise and falls in *"Twenty Years of Public Opinion about Global Warming."* More recent research can be found in the regular publications of the Yale Project on Climate Change Communication, especially its American Mind project. Models for segmenting public attitudes can be found in the regular research of Yale's *Six Americas project and the *American Climate Attitudes reports by the Social Capital Project.

Climate Communications

On the general communication of climate change, the *Climate Crossroads guide by Climate Access is a good introduction. There are a good range of online materials at www.talkingclimate.org and www.climatenexus.org. *From Hot Air to Happy Endings by Green Alliance is a collection of essays from leading theorists.

In writing this book I was especially intrigued by Judith Williamson's masterful analysis of the semiotics of climate change in her lecture *"Unfreezing the Truth: Knowledge and Denial in Climate Change Imagery" and Gill Eraut's examination of climate change narratives in *Warm Words: How Are We Telling the Climate Story?

Other Works by George Marshall and COIN

My first book, Carbon Detox (2007, Octopus), presents some of my ideas in this book in a digestible self-help format. The themes are further explored in technical reports by my organization, the Climate Outreach Information Network. Our website, www.talkingclimate.org, includes briefings on social norms and networks, behavior change, and science communication. Recent reports include *Climate Silence (and How to Break It) and *After the Floods: Communicating Climate Change around Extreme Weather. Regular reports are posted at www.climateoutreach.org .uk/resources/.

FOLLOW THE DISCUSSION AT
WWW.CLIMATECONVICTION.ORG

This book is designed to encourage a wide-ranging and long-running discussion about why we find it so hard to accept climate change and how we can build conviction. This is a lively and sometimes heated topic on which, it seems, everyone has an opinion. So please join in the debate and contribute your own thoughts and experiences and, especially, talk about it with the people around you.

On the book website at www.climateconviction .org you will find the line-by-line references and sources (regularly updated to include the latest research and opinions), briefing materials, reviews, and upcoming events.

You can join in the debate:
On Twitter @climategeorge
On the YouTube channel climategeorge
On Facebook at Don't Even Think About It—How Our Brains Ignore Climate Change
And on my blog www.climatedenial.org

And better still, download the PowerPoint presentation and speaker notes at www.climate conviction.org, and lead your own discussion in your group, workplace, college, or book club. We need to break this climate silence—and you are the person to do it!

Acknowledgments

Firstly I would like to thank everyone who gave up their time to be interviewed and so freely shared their insights and experience with me. You all influenced this book even if I have not been able to quote all of you directly. I hope that I have reflected or reported your views accurately and fairly:

Abie Philbin-Bowman, Adam Hochschild, Alya Haq, Andrew Simms, Annie Leonard, Anthony Leiserowitz, Ara Norenzayan, Atossa Soltani, Betsy Taylor, Bill Blakemore, Bill McKibben, Bob Buzzanco, Bob Inglis, Bruce Dobkowski, Bryan George, Cara Pike, Chris Rapley, Clive Hamilton, Courtney St. John, Cyndi Wright, Dan Gilbert, Dan Kahan, Daniel Kahneman, David Buckland, David Hone, David Reiner, Debra Medina and the members of the Texan Tea Party, Deyan Sudjic, Dina Long, Eamon Ryan, Erin Biviano, Erin Taylor, Eviatar Zerubavel, Frank Bain, Frank Maisano, Frederic Luskin, Gavin Schmidt, George Loewenstein, Gill Ereaut, Ginny Pickering, James Hansen, Jennifer Morgan, Jennifer Walker, Jeremy Leggett, Jim Riccio, Joanna Macy, Joe Romm, Joel Hunter, John Adams, John Ashton, John Charmley, Juliet Schor, Kalee Kreider, Kert Davies, Kevin Anderson, Kevin Wall, Keya Chatterjee, Kirk Johnson, Laura Storm, Marc Morano, Martin Bursik, Matthew Nisbet, Michael Brune, Michael Dobkowski, Michael Mann, Michael Marx, Michael Salvato, Myron Ebell, Oliver Bernstein, Oliver Tickell, Patrick Reinsborough, Paul Slovic, Peter Kuper, Renee Lertzman, Richard Muller, Ross Gelbspan, Sabine Marx, Sally Bingham, Sam Kazman, Sandy Dunlop, Stephen Lea, Steve Kretzmann, Ted Nordhaus, Thomas Schelling, Tim Nicholson, Tom Athanasiou, Vinay Gupta, Wendy Escobar.

I am immensely grateful for the support of so many others in the course of this project and my travels—people who have advised me on contacts and contracts, shared ideas and insights, put me up, and put up with me! Thank you again for your generosity.

Alastair McIntosh, Amara Levy-Moore, Andy Croft, Andy Revkin, Annie Leonard, Beth Conover, Bill Day, Brian Tokar, Bruce Rich, Bruce Stanley, Caroline Crumpacker and Roberto Rossi, Caspar Henderson, Chris Shaw, Claire Roberts, Clare Ellis, Cliff Jordan, Daphne Wysham, David Partner, David Rothenberg, Edward C. Chang, Erik Fyfe, Eve Levy, Fred Pearce, the Garrison Institute, Garson O'Toole, Geri and the staff of the Great Oak Café, Graham Lawton, Heart Phoenix, James Marriott, Jennifer Callahan, Jenny Hok, Jo Hamilton, Joanna Kempner, John Fousek, John Passacantando, John Seed, Jonathan Bunt, Jonathan Spottiswoode, Karen Hadden, Kari Norgaard, Katherine Hayhoe, Kathy Geraghty-Acosta, Lafcadio Cortesi, Lawrence Culver, Lisa Orr and Matt Harris, Lorraine Whitmarsh, Louisa Terrell, Lynn Englum, Mark and Jan Scott, Mark Haddon, Mark Levene, Mark Lynas, Matthew Jacobson, Michael Corballis, Mick West, Mike Roselle, Nick Lunch, Nick Pidgeon, Nicola Baird and Pete May, Olga Roberts, Paddy McCully, Pam Wellner and Eugene Dickey, Peter Lipman, Peter Winters, Rebecca Henderson, Randy Hayes, Richard Cizik, Richard Harris, Richard Hering, Robert Wilkinson, Ryan Rittenhouse, Saffron O'Neill, Sally Weintrobe, Sara Smith, Sarah Woods, Scott Parkin, John Houghton, Stephan Lewandowsky, Stuart Capstick, Susanne Breitkopf, Ted Glicke, The Authors Guild, Tom Crompton, Tom "Smitty" Smith, Tzeporah Berman, Zoe Broughton, Zoe Leviston.

My greatest thanks are due to the friends who peer-reviewed my rambling drafts and corrected my mistakes: George Monbiot, Hugh Warwick, Patrick Anderson, Paul Chatterton, Roman Krznaric, Vivienne Simon, and my colleagues at COIN, especially Adam Corner, Olga Roberts, and Jamie Clarke. Above all, though, my thanks are due to Dan Miller and Jay Griffiths, who provided meticulous line-by-line comments and proposed detailed changes. If this book is enjoyable to read it is a credit to their skills and knowledge.

Finally a big thank you to Nancy Miller at Bloomsbury who took a chance on an unknown English writer to write a book for the U.S. market on a subject that no one wants to read about: now that's brave publishing. And, always, my thanks and love to my ever supportive wife, Annie, and my children, Elsa and Ned, with my apologies for being a stressed and grumpy father.

Index

A NOTE ON THE AUTHOR

George Marshall is the co-founder of the Climate Outreach and Information Network, the first British nonprofit organization to specialize in public communication around climate change. COIN's international reputation is built on its commitment to reaching new audiences, including its pioneering communications work with trade unions, human rights organizations, and political conservatives.

Over the past twenty-five years George has worked at all levels of environmental and social rights campaigning: from direct action protests to governmental policy consultancies, with senior positions at Greenpeace USA and the Rainforest Foundation in between.

He is also the author of *Carbon Detox* (www.carbondetox .org), a slow-selling book offering "fresh ways to think about personal action to climate change," and he blogs every so often at www.climatedenial.org. He lives in mid-Wales with his family and several thousand comic books.